METHANE
Fuel for the Future

METHANE
Fuel for the Future

Edited by
PATRICK McGEER
Minister of Universities, Science, and Communications
Province of British Columbia, Canada

and
ENOCH DURBIN
Princeton University
Princeton, New Jersey

PLENUM PRESS • NEW YORK AND LONDON

Library of Congress Cataloging in Publication Data

Main entry under title:

Methane: fuel for the future.

"Proceedings of a world conference entitled "Methane—Fuel for the Future," sponsored by the government of British Columbia, held during September 1981, at Delta's River Inn, Vancouver, British Columbia"—T.p. verso.

1. Methane—Congresses. I. McGeer, Patrick L. II. Durbin, Enoch.

TP761.M4M469 1982 665.7 82-13120

ISBN 0-306-41122-9

Proceedings of the World Conference entitled "Methane—Fuel for the Future," sponsored by the government of British Columbia, held during September 1981, at Delta's River Inn, Vancouver, British Columbia

A Division of Plenum Publishing Corporation
233 Spring Street, New York, N.Y. 10013

Printed in the United States of America

PREFACE

In September of 1981, a world conference on alternative fuels entitled "Methane - Fuel for the Future" was held at Delta's River Inn in Vancouver, British Columbia. Approximately 500 registrants from over a dozen countries attended the two day meeting. There were 20 invited papers which form the basis of this volume. The conference itself was inspired by the "energy crisis". This crisis was not seen in terms of any real shortage of oil in the near term, although an end to conventional oil could be seen on the horizon. Rather, it was perceived as an artificial crisis, precipitated by OPEC, but one which required urgent and effective solutions.

Not everyone will agree that urgent action is required to meet the "energy crisis". Indeed, as this volume goes to press, the media are advising that a global glut of oil exists and that price reductions will inevitably ensue. The OPEC production rate has slipped from 31 million barrels a day shortly before the 1973 oil embargo, to a current rate of less than 20 million barrels a day. The non-Communist world now depends upon OPEC for less than half of its oil requirements versus 70 per cent only a decade ago.

Much of this reduced demand for OPEC oil is attributable to lower consumption from buyer resistance. Some is a result of alternative supply from major new oil fields such as the North Sea or Prudhoe Bay. But very little represents a shift to alternative fuels. That is unfortunate, because it is only through this latter method that true independence from OPEC can be achieved. It is also the only way a long range solution to the oil problem can be developed.

The world is just as dependent as it was a decade ago on oil. While new sources and a plateau in consumption have temporarily reduced OPEC leverage, the fundamental vulnerability to reduced supply remains and the punishing economic consequences of current prices continue.

Sheik Yamani of Saudi Arabia has perceived the world situation well and has articulated the consequences of various courses of action better than any Western leader. He has spoken frankly of the "oil weapon", and has cautioned his OPEC colleagues not to push the cost of oil to the point where Western nations will seriously pursue alternatives. So far the western world is acting precisely as Sheik Yamani would wish. Alternative energy projects are being abandoned as word of a glut of oil spreads, and as hopes for price reductions grow.

Saudi Arabia retains the greatest ability of all nations to influence the supply and price of oil because of its unequalled capacity to expand or restrict production. It can readily shift from a high of more than 10 million barrels a day to any figure below that. Thus, Saudi Arabia alone, through internal or external pressures, can produce an unmanageable glut of oil in the world, or leave it desperately short. Without Saudi Arabia production there is simply not enough oil to go around.

For the first dozen years of its existence, OPEC had virtually no success in escalating oil prices. It was a chance accident to the trans-Arabian pipeline in May of 1970 that touched off the first major supply shortage. Only then was the true inflexibility of Western demand demonstrated, and the route to OPEC power clearly outlined. The price shocks of the 1970s, each tied to OPEC supply restrictions, are subsequent history. Either the price was paid, or homes could not be heated, electricity could not be generated, and vehicles could not move. With no alternatives to OPEC oil, the choices facing Western consumers were economic dislocation due to lack of fuel or economic dislocation due to higher prices. Higher prices proved to be the course of minimum regret although restricted supply, with its more severe consequences, could be imposed at any time.

These lessons of history need never be repeated if the course of action outlined in this volume is followed. Oil must eventually be replaced as an energy source, and particularly as a transportation fuel. It can be done now, in which case dependence on OPEC will be terminated, or it can be done later, in which case dependence on OPEC will continue. The choices for replacement fuel are few. They include methane, hydrogen, methanol, ethanol and synthetic fuels from coal. For many countries methane is by far the most attractive choice, both in the short and long term.

The reasons for methane's superiority -- availability, price, ease of distribution, cleanliness and compatibility with current engines -- are outlined in the early chapters of the book. It is the aspect of compatibility which may be the most intriguing feature of methane as a transportation fuel. Vehicles can have a dual fuel capability, in which case they can utilize OPEC-derived gasoline or diesel if price and supply are attractive, but can readily switch to domestically-produced methane if they are not.

The state of the art technology with respect to methane utilization in vehicles, described in the middle chapters of the book, provides insight into how rapidly a wide-spread system could be introduced into a jurisdiction such as North America. The time line could be measured in months as opposed to the years required for the development of synthetic fuels programs or major new oil fields.

The political problems associated with alternative fuels programs, which form the final chapters of the book, deserve to be particularly closely studied by policy makers. It is never easy to effect major changes in industry. Revolutionary ideas precipitate change, but the industrial development which follows is usually evolutionary with the logic of economics dominating decision making.

The energy crisis is different. The price and supply of oil are unrelated to oil reserves, production capacity, or rational economics. They are related to politics, primarily those of the unstable Middle East. Uncertainty of price and supply is the greatest weapon OPEC can use to keep the oil importing nations in bondage and they have used it well. Prices and supply have risen and fallen unpredictably. The response of Western nations at the political level has been one of confusion and uncertainty, thus the logic of internal economic decision-making has been impossible to apply. Firm policies that could be translated into appropriate economics have never really evolved. But the closing chapters reveal the sensitivity of response to the details of various government alternative fuel initiatives. It is here that the volume may have its greatest value because it provides evidence that the commodity most lacking today is not fuel but leadership.

ACKNOWLEDGEMENTS

The editors are grateful to a number of individuals whose dedicated efforts were essential to the production of this volume. Foremost amongst them is Jane Burnes, Policy Coordinator in the Ministry of Universities, Science and Communications of the Province of British Columbia. She undertook the task of assembling and coordinating all of the written material, supervising the artwork and preparing the index. Typing of the volume was done by Mrs. Judy Fetterly, also of the Ministry of Universities, Science and Communications.

The volume represents the transactions of the world conference "Methane - Fuel for the Future" held in Vancouver, British Columbia in September of 1981. The conference itself was co-sponsored by the Energy Development Agency and the Ministry of Universities, Science and Communications of the Government of British Columbia. Organizers of the conference, in addition to the editors, were Mr. Ray Langdon, Dr. R.W. Stewart, Mr. E. Macgregor and Mr. George Lenko. Special help at the conference was provided by many others, too numerous to list.

We are indebted to the following firms who presented exhibits of the state of the art technology at the conference:

- ADVANCED FUELS, Wichita, Kansas
- BEECH AIRCRAFT CORPORATION, Boulder, Colorado
- BLACK TOP CABS, Vancouver, British Columbia
- B.C. HYDRO, Vancouver, British Columbia
- CANADIAN WESTERN NATURAL GAS, Calgary, Alberta
- CNG FUEL SYSTEMS, Calgary, Alberta
- DUAL FUEL SYSTEMS, Culver City, California
- EUROMATIC MACHINE AND OIL COMPANY, Jersey, England
- FIBA-CANNING, Toronto, Ontario
- GAS SERVICE ENERGY CORPORATION, Kansas City, Missouri
- J-MAR, Richmond, British Columbia

- L.P.F. CARBURETOR, Hutchinson, Kansas
- NORWALK, South Norwalk, Connecticut
- RIX INC., Emeryville, California
- SPRINT AUTO, Bologna, Italy
- STRUCTURAL COMPOSITES INDUSTRIES, INC., Azusa, California
- TRI FUELS INC., Plano, Texas
- WILLIAMS & JAMES, Gloucester, England

CONTENTS

D. COMMERCIAL EXPERIENCE WITH ALTERNATIVE FUEL PROGRAMS

E. SOME NATIONAL PROGRAMS FOR ALTERNATIVE FUELS

THE URGENCY FOR MULTI-NATIONAL ALTERNATIVE FUELS PROGRAM

THE URGENCY FOR A MULTI-NATIONAL

ALTERNATIVE FUELS PROGRAM

Patrick L. McGeer, Minister of Universities, Science and Communications, Government of British Columbia
and
Enoch J. Durbin, Professor of Mechanical Engineering
Princeton University

In 1973, the OPEC nations served notice upon the rest of the world that they no longer intended to supply oil at what they considered to be the bargain basement price of $2.50 a barrel. They said they were going to increase prices and that western exploitation of the vast fields of the Middle East had to be reduced.

The OPEC nations have certainly delivered on their warning. World prices have increased 15-fold since that declaration. While the production of oil is still adequate to meet world needs, allocation is uncertain and anxiety prevails. The Middle East is one of the least stable political areas in the world and the United States and Canada are dependent upon it for oil. The foreign and domestic policies of every oil-short country in the world have been re-assessed and compromised as each country tries to ensure its security of supply and to cope with the outflow of its domestic funds. The OPEC nations, adjusting to their vast new riches, have often responded in unpredictable ways to the pressures of world power politics.

Businessmen and consumers in the private sector of western nations are confused as to what the future holds in energy matters.

It would be hard to over-estimate the impact of these changes. Total payments to the oil exporting nations from oil importing nations runs at a rate approaching one billion dollars a day. The consequences have been many. Transfer payments have been too overwhelming for OPEC nations to absorb, and too large for donor nations to maintain. One result is that world currencies have been universally debased. The massive recycling from the Middle East of key western currencies which underpin the world's monetary system, particularly the American dollar, are now outside of the national controls of the countries of origin. Money supply can no longer be controlled by their central banks. The full consequences of this situation are yet to be determined, but continuing inflation seems certain to be one.

For Third World nations, the consequences of the oil price increases have neen nothing short of tragic. Unable to power their machinery, they sink farther behind the west, accumulating political problems for the future. The OPEC pricing initiatives and production policies, while adversely affecting the world's economy, have to date stopped short of driving consumer nations into an all-out effort to find substitutes for oil as a transportation fuel. Sheik Yamani of Saudi Arabia has stated that this is indeed the strategy that OPEC must employ. Nevertheless, this replacement policy should be the top priority of all consuming nations. Only in this way can the transportation symbiosis that has gradually evolved between consumers, oil companies and auto manufacturers be re-established and the world's economy restored to better order. The oil substitutes must be cheap enough to provide competition for OPEC oil. Otherwise they will just be a spur for even further oil price increases.

The Effect of OPEC Price Increases On Individuals and Companies

The United States, currently spending about $85 billion per year on OPEC oil, has adopted policies to repatriate significant portions of its currency outflow. The approaches it has implemented of high United States domestic interest rates and arms sales to the Middle East have greatly worried other nations. The resulting distortions of the domestic economy have been severe. In the North American automotive industry, for example, consumers are withholding new car purchases. High interest rates are an important contributing factor. Sales in the third quarter of 1981 were the worst in modern history. Housing starts in this period were the lowest in 40 years. The layoffs which have resulted have produced a recession spreading over many other sectors. Thus, what has been gained in repatriated foreign currency must be weighed against what has been lost in the distortion of the entire economy.

Since most oil is consumed in transportation, a significant proportion of the transfer payments to OPEC nations comes indirectly from individual consumers operating their own vehicles. The size of the payments has obviously shifted the habits and desires of these consumers, which has had its own massive effect on the economics of western nations.

Over the past three-quarters of a century, a stable symbiotic relationship had developed between the consuming public, car manufacturers, oil companies and governments. It had evolved so gradually that few had even become aware of its existence. Yet its components are now so large as to dominate the economic well-being of many leading nations. Of the eight largest public corporations in the world, in terms of sales, five are oil companies and two are automobile manufacturers. The United States automobile related industry by itself accounts for one-seventh of America's gross national product. It is responsible for one-seventh of all U.S. employment, consumes one-fifth of the steel output, one-quarter of the glass output, one half of the malleable iron output, and three-fifths of the synthetic rubber production. Over 140 million vehicles travel on American roads, consuming one barrel in ten of the world's oil production.

The oil companies provide the fuel for cars built by the automobile companies, but both are dependent for sales on individual consumers. Governments build and maintain the roads for consumers to drive their cars on, while deriving signficant revenues from various taxes on both automobiles and fuels. Although they depend on votes rather than sales for their success, they too must anticipate the desires of consumers. The interrelationships are roughly as shown in the diagram below.

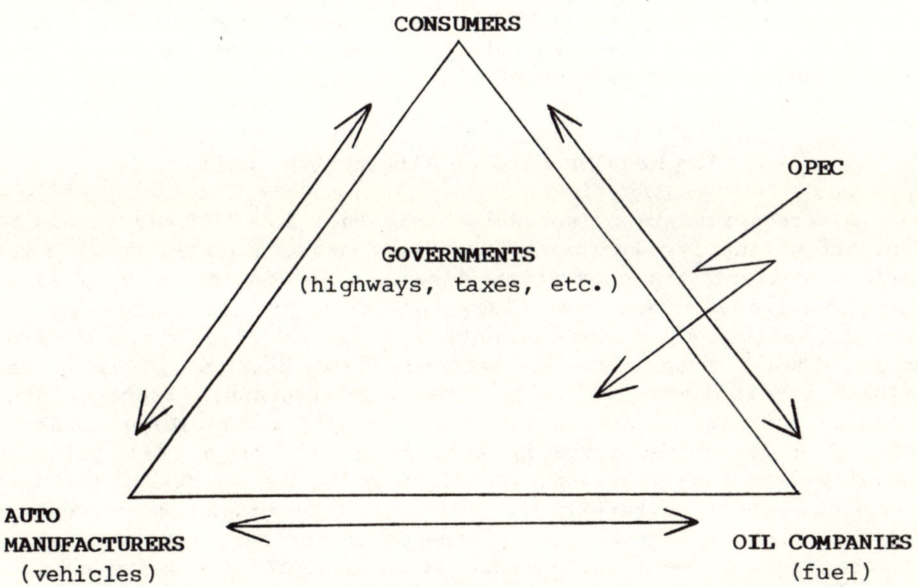

FIGURE 1. Symbiotic Relationship of Fuel Companies, Auto Manufacturers, Consumers and Governments

The diagram indicates the nature of the stable symbiotic relationship which has now been unbalanced. A key figure at the apex of the triangle is the consumer. The cumulative effect of his decisions is critical to the economic survival of automobile and fuel companies, and the political survival of governments. OPEC, by raising its price of oil to fuel companies, has increased the cost to consumers and, perhaps more importantly, raised questions about its availability. This has disturbed the stability of the symbiotic relationship depicted in the diagram. There is a shift in the types of cars people will buy and in their driving patterns. What further shifts might take place cannot be forecast. The predictability of consumer purchasing and driving habits, which is so essential to rational economic planning, is subject to much greater uncertainty. It had been based on slow evolution of the symbiotic relationship and mutual confidence among consumers, governments, fuel companies and car manufacturers.

But the consumer, in addition to modifying past purchasing patterns, no longer trusts governments on energy matters. This is not limited to OPEC. It also applies to domestic governments that have spawned energy ministries in response to OPEC. Those ministries have developed a myriad of complex policies around the world. The consuming public is getting mixed and confusing signals about prices and availability. The consumer understandably wants different types of cars because of fuel uncertainties. Automobile companies are trying to identify what market opportunities exist but this, in turn, often depends on domestic energy policies which are subject to short-term change.

The Requirements of Alternative Fuels

The requirements of any substitute fuel for OPEC oil should be a plentiful supply, a competitive price, and compatibility with the world's current transportation fleet. The fuels which fulfill these requirements are few. They include major new sources of oil from conventional or non-conventional fields, liquid fuels from coal, ethanol from biomass, methanol from coal or methane, and methane itself from gas fields, coal, and biomass. Propane, proposed by some as an alternative fuel, is only a very minor constituent of a gas field (usually less than 5%), or an oil refinery (usually less than 10%), and is therefore in too short supply to be a useful general alternative. Further, the price of propane is volatile since its availability depends on refinery runs and natural gas usage, as well as demand. It is nevertheless being used on a large scale in the gas-rich Netherlands and on a smaller scale in Japan and North America. Current gasoline engines can be readily adapted for use of propane. Most propane fueled vehicles are not dual fuel.

New sources of oil from either conventional fields, or such non-conventional sources as tar sands or shale oil plants will not only be much higher priced than today's OPEC oil, but will require up to a decade for building the necessary plants and infrastructure after decisions have been made to develop production from this source. The availability of such fuels will automatically raise the asking price of conventional oils.

Liquid fuels from coal are not a viable alternative in the short to medium term for the same reason. Although Hitler's tanks and Messerschmidts ran on liquid fuel made from coal, that technology has advanced little since World War II. Immediately after the war, Germany was forbidden to rebuild its coal liquefaction plants. By the time the embargo was lifted, low priced Middle East oil was flooding European markets and the technology slipped into limbo. Only isolated South Africa pursued coal liquefaction and today their SASOL plants are the only major facilities in the world capable of producing liquid fuels from coal. Some small coal liquefaction demonstration plants are in operation in the United States, but at least 10 years will be required for this alternative to have any substantial impact on the need for gasoline and diesel fuels obtained from conventional oil. Current estimates place the price considerably higher than current OPEC prices.

More serious alternatives are ethanol from biomass, methanol from coal or natural gas, and natural gas itself. Each can be made available in sufficient quantities to displace OPEC oil. While costs will vary from country to country, in North America the current relative production costs can be readily estimated since each of these materials is being produced in large industrial quantities. For 100,000 Btu's (approximately 0.87 of the United States gallon of gasoline) the consumer will probably pay at the service station pump roughly 50 cents for compressed natural gas, $1.00 for gasoline, $1.15 for methanol, and $2.20 for ethanol exclusive of taxes. Obviously, such figures are approximate and subject to geographical variation even without the impact of regional taxes.

Brazil is currently producing ethanol from sugar cane on a scale sufficient to operate a significant proportion of its transportation fleet, as described in detail in Chapter 20, by Drs. G. Pischinger and R. Siekmann. That approach is obviously a viable alternative for Brazil chiefly because of its labor intensive agriculture. For North America, the disadvantage is the substantially higher cost of alcohol as opposed to gasoline at today's OPEC prices. Alcohol can be used as a fuel dilutent, gasohol, which is gasoline containing up to 20% anhydrous ethanol. It has been shown that ethanol from agricultural crops in non-labor intensive agriculture consumes more oil energy than it displaces. In North America there is a net oil energy cost of about 30%.

Methanol is currently being produced on a large scale from methane at about 65% efficiency. It can also be produced from coal as described by Dr. R. McElroy in Chapter 6. So far automobile engines have not been produced to utilize this fuel but presumably the technology would be comparable to that for ethanol.

The alcohols, then, are higher priced and when based on agricultural products are of no value in relieving dependence on oil imports. They are incompatible with today's gasoline and diesel engines unless used in their anhydrous state as minor gasoline dilutents. As Brazil has demonstrated, alcohol burning engines are quite feasible.

This leaves methane, obtained either from gas wells or coal, as the remaining serious alternative for nations like the United States and Canada which have generous supplies of the raw material.

Methane as the Viable Alternative

What are the characteristics of methane as an engine fuel? Methane has an octane number of 130, far above that of the 89 octane gasoline which is burned in our vehicles today. The efficiency with which one can use a fuel in an engine depends on the octane number. Methane is environmentally superior to gasoline. It is so clean, safe, and non-toxic, that we do not hesitate to burn it in our kitchens for cooking or in our basements for heating our homes. Its use would therefore tremendously reduce the smog that occurs in our large cities, due to automobile exhaust, without the use of expensive catalytic converters or other clean-up devices. It prolongs the life of automotive engines by not diluting the lubricant. Since it is a gas, even at low temperatures, methane makes engine starting easier than with liquid fuel. A natural gas engine requires no choke, even in cold weather. It has a high ignition temperature and a narrow inflammability limit. Consequently it is a lesser fire or explosive hazard than gasoline. It is therefore safer in case of an accident, independently of the fact that it is normally carried in a thick-walled pressurized cylinder rather than a thin metal container as is gasoline. It is lighter than air and, unlike propane which puddles when released, it rises and quickly disperses when it escapes. Forty years of Italian experience have demonstrated its inherent safety as a motor fuel.

Over one million miles of pipeline deliver natural gas to the majority of homes in North America and could deliver it to almost every service station in the nation.

It has the tremendous advantages, then, of having a distribution system for it almost totally in place, of not requiring any further refining or chemical modification to be burned in existing internal combustion engines, and of being much cheaper than gasoline.

What are the disadvantages? There appear to be only two, apart from the cost of retrofitting current engines. The first is the limited fuel capacity and bulkiness of the high-pressure methane storage tanks. The second is a modest loss of maximum power in a methane powered engine. The loss occurs because gaseous fuel displaces some of the air intake, thus impairing the engine's air breathing capacity. A second loss occurs because methane is a slow burning fuel (hence its high octane rating). Some of this latter loss is diminished by use of advanced spark timing. In an engine adapted for use with natural gas by means of a higher compression ratio (12:1) or turbo charging, the loss of power is completely eliminated and the engine efficiency in the use of fuel is enhanced by at least 25%.

World Availability of Methane

Despite the impressive advantages of methane, efforts to use it have been so retarded that significant amounts simply go to waste. For instance, in 1980 alone, 7.5 trillion cubic feet of methane was flared by oil companies around the world: equivalent to more than 60 billion gallons of gasoline. Most of this is flared in the Middle East. Indeed, among the most prominent nocturnal objects viewed from satellites are not the lights of the world's great cities, but the gas flares of the Middle East. The fuel wasted by this profligate practice would be sufficient to power two-thirds of Europe's road transport.

In Indonesia, where the three oil fields of Badek, Nihem, and Handil flare 300 million cubic feet of methane every day, the folly is being corrected. A pipeline, which will collect the gas and carry it to an LNG production plant, is now being built. It will double exports of liquefied natural gas to Japan. In Saudi Arabia extensive chemical plants are being built to utilize this "waste gas".

The flaring figures are disturbing - but so are other figures, where methane is not wasted but shut-in. In Alberta, over 7500 wells are capped. In British Columbia, as in New Zealand, governments that offered take-or-pay contracts to producers in order to stimulate exploitation now have such a large surplus that huge amounts of money are being paid for gas that can't be used. Enormous amounts of gas are shut-in on Alaska's north slope. Although gas is often discovered when new sources of oil are being sought, there is no economic incentive to delineate the boundaries of a new field, and therefore no systematic approach exists for evaluating potential new supplies.

The truth is that there is a glut of methane in many areas, particularly in remote oil fields where the practice has been to flare the gas dissolved in oil that is being pumped. As a result, sources of methane other than conventional gas and oil fields have never been seriously explored. Methane is the major gaseous product from coal, peat, and shale oil. It is also the major gaseous constituent from anaerobic digestion of sewage and other biomass and it can be produced from water, solar or nuclear energy. If all fossil fuels ran out, which is unlikely for centuries, methane is the best renewable, safe, high-energy density fuel that can be made from combustion engines. It is better than hydrogen since the energy per cubic foot is more than three times greater and at the present time it is far, far cheaper.

Why Not Methane Now?

The question then is: why isn't methane being used now in the oil importing nations instead of gasoline? Of course it is, in about 400,000 vehicles around the world, but that is a drop in the bucket. The main reason is the existing world-wide commitment to gasoline and diesel fuels, which poses an economic threshold that is difficult to overcome. There is an existing infrastructure and business complex which is firmly entrenched.

This threshold exists for other fuels and has to be overcome at the individual level for any alternative fuels program to be effective. Oil short nations can only gain the benefits of independence from OPEC, retention of domestic earnings, reduced inflation, and a cleaner environment, if substantial numbers of vehicle owners make a personal decision to change from OPEC oil to the alternative fuel of choice. Each vehicle owner who makes a change must gain some advantage before the state, in turn, can gain. Therefore, one has to turn away from all the global considerations that have been discussed so far and focus right down on factors influencing the average vehicle owner in his consideration of the possibility of buying or equipping his car to use an alternative fuel.

The particular hurdle which inhibits car owners from utilizing methane as a second fuel, is that while the engine can burn methane, it has not been designed for such a purpose. Furthermore, the car body has not been designed to accommodate appropriate storage tanks for methane. It must be compressed about two hundredfold into a cylinder or liquefied by temperature reduction to -161 degrees Centigrade which concentrates it six hundred-fold. Therefore, the car owner faces substantial costs for the retrofitting of the engine and the addition of the cylinders.

The costs would be much less if these options were incorporated at the factory level, but no automobile manufacturer today provides a dual fuel opportunity. At retail service station prices the retrofit cost, including the storage tanks and labor, comes to more than $1,500 in North America. This entails 8 hours or more of labor and about $500 (wholesale) worth of components. Of considerable concern to the vehicle owner who contemplates conversion is his uncertainty as to cost differences between gasoline and natural gas for the future. Since most governments regulate gasoline and natural gas prices, this uncertainty represents a lack of confidence in predicting future government policy.

Since the gains to the state by conversion are immediate, and the gains to the individual are protracted, it is necessary that a government system of incentives be developed which will assure the individual of a satisfactory return on investment. This can take many forms. Assured price differentials are the single-most important incentive. By special grants, the individual can recover his investment over a shorter period by his lower fuel costs and lower engine wear. With small grants and small price differentials, conversion only becomes attractive to taxi or large fleet operators where vehicles put in very large mileages. In New Zealand, grants to individuals have produced substantial results in terms of conversion of private vehicles, as described in Chapter 16 by Peter Graham. The New Zealand government's target of 150,000 conversions, or 20% of the automobile fleet, by 1985 is likely to be easily achieved. Recently, Canada and the Province of British Columbia introduced conversion incentives to individuals in the private sector which can be expected to influence substantially the number of automobiles operating on methane in Canada.

The contribution of $600 from the Government of Canada plus $200 from the Government of British Columbia for each car converted, plus removal of the road tax on methane in British Columbia, should soon have a significant impact on alternative fuel usage.

A further hurdle is with respect to refueling. Unfortunately, an economical mini-compressor which would permit an individual to refuel at his own home is not yet on the market. Therefore, the car owner must find a service station prepared to supply the fuel. Service station owners, of course, are not going to establish refueling stations unless there are many cars in the vicinity that will use the station and therefore permit the operator to recover the cost of the compressor equipment. A 600 liter per hour (gasoline equivalent) station can cost as much as $150,000 in additional equipment. The operator is not going to make a decision to purchase this equipment unless he can see an appropriate financial gain from doing it. Once again the advantage accrues to the state immediately from the availability of a methane refueling station, but not to the individual service station operator.

In New Zealand a one year tax write-off of this equipment has been sufficiently attractive to develop a natural gas dispensing network.

Not to be overlooked are the direct benefits to gas companies. There is a significant overhead associated with operating the one million miles of pipeline that distribute natural gas to individual homes in North America. In Chapter 15, John Wright draws attention to the fact that there has been a reduction in consumption in the Kansas City service area from 210,000 cubic feet per customer per year in 1970 to 125,000 cubic feet per year in 1980. On the other hand, the amount of methane one of his company fleet cars uses in a year is roughly equivalent to 1.25 homes in the service area. Automobiles therefore are a significant way of improving the efficiency of the existing natural gas distribution system.

Conservation measures have resulted in many pipelines operating at less than capacity. Nationwide, the use of natural gas in the U.S. has dropped 20% in the 1970-1980 decade. During the summer season when home heating is at a minimum, car driving is at a maximum: thus, use of methane as an automobile fuel would nicely complement its use for home heating, and would allow pipeline overhead costs to be reduced. Economic analysis suggests it might even pay gas companies to supply home compressors for automobiles free of charge. The compressor cost is comparable to the cost of lines and meters to add a single new home to the system. Certainly a methane service station would supply a gas market equivalent to a large urban subdivision and at a greatly reduced overhead cost.

The financial difficulties of conversion are compounded by legal and regulatory hurdles. Fire and safety regulations and other bureaucratic decisions can delay decisive action. In the United States, a priority research area under the "Methane Transportation Research Development Demonstration Act of 1980" is the identification of institutional barriers to the use of methane. This would include such inappropriate regulations as the restriction of vehicles equipped for methane from using tunnels or ferries, or from driving across state boundaries if the methane sold in one state is prohibited from being used in a neighboring state.

The experience of many nations since the OPEC declaration of 1973 makes it seem obvious that the existing world-wide commitment to gasoline and diesel fuels, combined with the financial and institutional barriers to the introduction of alternatives, are too strong a combination for the private sector to overcome unaided. Firm government leadership is required to effect changes.

No national government seems to have come close to weighing accurately the economic and other advantages to be gained from using an alternative fuels program to eliminate dependency and to control OPEC oil prices permanently. For example, at current levels, the value of material and labor in a new car is less than the cost of oil-derived fuel that the car will consume over its normal lifetime. Each car in excess of the number that can be supported by domestic fuel represents a commitment to OPEC fuel. Therefore, each such car becomes a national liability in terms of foreign currency even if it is domestically produced. The additional benefits of political independence, reduced inflation and improved environment are intangible, but not beyond a government's ability to attempt assignment of value.

Such assignment seems to have been made, perhaps not intentionally, in the case of the U.S. environmental standards for automobile emissions. Automobile manufacturers reacted to legislation by producing platinum catalytic converters which are incompatible with leaded gas. The fuel companies had to abandon leaded fuel. They are now forced to operate their refineries at about 8% less efficiency in order to produce unleaded gas of the requisite octane number. Since this unleaded gas has a lower octane number, the automobile manufacturers must lower engine compression ratios. This has resulted in a family of engines which operate approximately 7 1/2% less efficiently than their predecessors with higher compression ratios. Thus, the cost of this environmental requirement is approximately 15% of the total gasoline used in the United States. It could be said that this incremental 15% is entirely imported, amounting to a cost of about $13 billion annually. It may never have been the intention of legislators to have industry choose a solution so costly to the nation in response to their regulations. It was, however, the least costly solution for the auto industry. A more thoughtfully constructed regulation could have resulted in a lower national cost.

The national benefits of natural gas as a fuel are so high that sharing a significant portion with individual vehicle owners should be sufficient to spur dramatic alternative fuel demand which could spur equipment manufacturers, and perhaps even automobile manufacturers, to make commitments to produce the appropriate hardware.

These commitments should incorporate technological advances over the rather ancient current equipment. What are these technological advances? Some of them are discussed in Chapter 7 by Enoch Durbin, in Chapter 8 by Durbin and Gerard Born, and in Chapter 10 by Robert Axworthy. For the engine itself, reliable ignition and timing devices need to be developed which can accommodate a diversity of fuels. A stable delivery and metering system for methane should replace the unstable low pressure feed system of today's kits.

Perhaps most importantly, advances need to be made in fuel storage systems. The compressed methane cylinders are bulky, costly, and permit a range of less than one-half that of the average gasoline tank on today's vehicles. Better storage technology with a larger fuel capacity is badly needed. Lower cost storage devices are desireable. Storage of methane through absorption on a surface could increase the capacity and reduce the required pressure. That would represent an enormous advance. Better cryogenic systems might permit liquid methane, as opposed to compressed methane, to be used as the storage fuel. The development of a low cost compressor for home-use would permit individuals to attach this device to their home gas line and to refuel overnight at home without a service station. Methods for metering gas for accurate billing at dispensing stations need to be developed.

While great advances can be made, it should be emphasized that acceptable systems are available today making all of these suggestions merely rich, added benefits. It should also be emphasized that, for the foreseeable future, automobiles should be equipped for dual fuel. The gasoline option should not be dropped unless and until a highly developed and sophisticated methane refueling system is universally available. But that does not mean that engines of the future should continue to be designed primarily for gasoline with the option of shifting to methane. The reverse should be true. It is far more satisfactory to design a high compression ratio engine for methane and adapt it down to use gasoline than to do the opposite.

Are Long-Term Supplies Adequate to Justify Commitment to Methane Technology?

If the methane option is to be seriously pursued, citizens of the United States are bound to ask about the long-term adequacy of supply. U.S. reserves of natural gas and oil in equivalent energy are about equal, but natural gas use is about one-half that of oil use. With proven gas reserves now estimated to be enough for about 25 years, at current production rates, what of the future? As Jack Bowen points out in Chapter 3, it is illogical from a financial point of view to drill for natural gas that has no prospect of being used. Therefore, demand paces supply. Proven reserves reflect financial policy more than geological reality. Nonetheless, a long-term commitment to methane as a fuel requires a more certain backup. This security is dealt with in Chapters 4, 5, and 6 by Drs. Gold, Antal and McElroy. It is particularly noteworthy that coal can ensure methane supplies for centuries. Favourable deposits can do that at a cost which is apparently below some sources of new conventional gas in North America. Thus, there need be no fear of supply shortage and no fear that costs will be out of proportion to the accepted rates of today. Natural gas can indeed provide a restraint on the price of imported oil.

What the United States should do if it were to pursue the methane option aggressively is to use the several decades granted by the known methane reserves to develop, as a fail safe source for the future, a capability of making high Btu gas from coal fields or even biomass and feeding this into the existing pipeline system.

For a few nations, such as Canada and New Zealand, this discussion need not even take place. Supplies are already enormous, to the point where new exploration is not being vigorously pursued because the proven supplies cannot be consumed in the immediate future.

The Need for a Global Strategy

What is presently lacking is a firm global strategy for implementing a changeover to methane. It is not a substitute fuel, it is a superior fuel. But the changeover cannot be achieved unilaterally by automobile companies because they lack the capacity to make fuel available. It cannot be done unilaterally by oil or gas companies because they do not make the automobiles or the equipment. It cannot be done unilaterally by the governments of most countries because they are not large enough to influence the international automobile, oil, and equipment companies.

All of this implies that nothing short of an international, or global, strategy to develop the methane alternative will be adequate. Politicians, automobile companies, equipment manufacturers; oil, gas and coal companies, scientists and engineers must work together.

What should this global strategy include? The cardinal element is to recognize that incentives must be provided to the motorist. The state must be prepared to transfer some of the benefits it gains to the individual who converts in order to obtain the political rewards of genuine energy independence and fiscal stability.

In addition, further incentives must be provided to distributors to stimulate the development of a comprehensive system of refueling stations so that owners can use this fuel anywhere.

Next, the automobile manufacturers need to be given a stronger reason than they now have for investing in the technology of improved methane burning engines and storage systems. This could be in the form of reduced taxes on newly manufactured automobiles in this category and subsidies for retrofit kits on existing engines.

Global cooperation will be required to produce comparable financial incentives from nation to nation, to provide for the exchange of data for appropriate safety and technical standards, and to share in research and development. Finally, research and development must begin to exploit the methane sources beyond conventional gas fields.

As Congressman Dan Glickman's testimony in November, 1981 before the U.S. House of Representatives Science and Technology Committee on Transportation, Aviation and Materials indicates (see also Chapter 21), such moves are likely to be discouraged by OPEC. Sheik Yamani of Saudi Arabia has warned his country that it is not in Saudi Arabia's interest to raise oil prices to the point where the United States would seriously pursue alternatives to OPEC oil. Not only would Saudi Arabia lose its most important market, its power to influence U.S. policy would disappear.

To quote Sheik Yamani's statements in Jidda in the fall of 1981:

> "As a result of the Saudi production and pricing policy many major companies have been very reluctant to implement their energy substitution projects. This is in the interest of the Arab cause in that it restores the importance of oil.....Oil as a political weapon will come back once again when there is a balance between supply and demand."

So far, OPEC strategy has worked perfectly. Prices have been increased to a point where the transfer of wealth beggars any previous transfer in world history. Yet prices have not been so high as to persuade western nations that they must aggressively pursue alternatives. Nor have the OPEC nations harmed in any financial way the giant international oil companies. Their sales and profits have soared to record highs. Automobile companies, particularly in the United States and Canada, have suffered but it is not clear that this results directly from OPEC prices, nor is it clear that long-term recovery depends on secure alternative fuels that have a predictable price.

Subsequent chapters in this volume will present the detailed arguments for an alternative fuels program now. They will advance in greater detail why methane is not only the most viable substitute, but is also a superior fuel in its own right. Various nations can draw heavily on the extensive experience with alternative fuel programs that has already been gained in Italy, the Netherlands, New Zealand, and Brazil. These are presented in

Chapters 16 through 20. The lesser experience of the United States and Canada is also described. Other chapters are intended to cover in some depth the current state of the art with respect to engine technology and dispensing of methane and other alternative fuels.

We hope that you, the reader, will find a niche in this endeavor where you might participate.

AVAILABILITY OF ALTERNATIVE FUELS

GEOLOGICAL ESTIMATES OF METHANE AVAILABILITY

Ken Deffeyes

Professor of Geology

Princeton University

The product that we know as natural gas is not pure methane. A few percent of molecules with two and more carbon atoms are usually present. The molecules with 3, 4, and 5 carbon atoms are variously referred to as natural gas liquids, gas condensate, or casinghead gas. Because these molecules with more than one carbon atom are likely to be in demand as petrochemical feedstocks and as liquified natural gas, we can presume that they will normally be removed from natural gas streams and marketed separately at higher prices.

Geological estimates of the reserves of natural gas on a country-by-country basis are not presently available. However, if there were accurate estimates of the total resources of worldwide natural gas, there would be no purpose in holding a conference on the future of methane as a fuel. The reason is that production has habitually lagged behind the additions to reserves by about 12 years. Those reserves which are already identified are a partial basis for optimism about the expanded use of methane as a fuel. However, since many of the implementations discussed in other chapters in this book will take 5 to 10 years to reach maturity, much of the natural gas to be used in those programs is gas that has yet to be discovered. In many instances, the regions are known to contain gas and individual fields have been discovered, but the detailed development drilling which would justify specific estimates of reserves will only be available a few years before the methane is actually put into the pipeline.

Because of this time lag between establishment of reserves and production, the most relevant question for today is the discoverability of additional natural gas. The decreasing rate at which oil is found, initially in the well-explored countries and now for the world, has brought on the so-called "energy crisis". Since natural gas has traditionally been found by the same exploration programs as oil, one might question whether the discovery of new gas reserves today is as difficult as finding new oil. There are several reasons why the exploration cycle for natural gas is at an earlier stage than the cycle for oil, and it is the purpose of this paper to explain why that is true.

The key to the discussion is recognizing that there are some terrains which produce natural gas that are quite barren of oil. It is these terrains, which oil exploration programs learned to avoid (through painful and hard knocks), that are the drilling targets newly opened by increasing demand and prices for natural gas.

There is a similar lesson contained in the experience with uranium. Initially, the amount of low-grade uranium ore was estimated by taking the ratio of high-grade ore to low-grade ore in existing uranium mines. Because these estimates showed a very small amount of low-grade uranium ores, a program of developing the breeder reactor made economic sense. However, once those low-grade ores which were not directly associated with high-grade ore were included, the amounts of uranium available at moderate cost were so large as to eliminate the economic need for the breeder reactor. To avoid making the same mistake for oil and gas, we need to be careful to examine separately the methane source associated with oil and non-oil bearing methane sources.

Natural Gas Associated With Oil

Natural gas that is found in direct association with oil occurs in two forms. Even where the oil is under-saturated with natural gas, a considerable amount of gas is typically dissolved in the oil under the initial reservoir conditions before production begins. This gas is separated from the oil as it is produced: much of it is still disposed by burning in large gas flares near the producing wells. Where a market exists for the gas and where pipeline connections exist, the solution gas can be sent to market.

A second, and different circumstance occurs where the reservoir contains more natural gas than can be dissolved in the oil. Then the gas segregates within the producing formation into a gas cap. Even where no gas cap exists naturally, sound petroleum production practice often includes the creation of an artificial gas cap.

There are several reasons for establishing an expanding gas cap:

1. The percentage of the oil swept out by the expanding gas cap is a higher percentage than would be recovered from the direct expansion of the gas in solution in the oil.

2. Maintaining the reservoir pressure as high as possible through gas reinjection keeps the gas bubbles within the oil-bearing formation as small as possible, thereby increasing the relative permeability for oil.

3. After the oil has been recovered, the gas cap itself can be produced and marketed.

Because of the economic and conservation reasons for reinjecting solution gas into oil reservoirs, it is truly astonishing that the practice of flaring natural gas continues on a large scale, particularly in North Africa and the Middle East.

As recently as 30 years ago, solution gas produced with oil could supply all of the consumer market for gas in most countries. Prices for gas were typically 3 to 9 cents (US) per thousand cubic feet at the wellhead. Under those conditions, the production and distribution of natural gas became regulated as a utility, rather like the distribution of water and electricity. As long as most of the natural gas was produced from association with oil, it was rather easy to estimate the amounts of gas that were expected from existing wells.

Natural Gas Not Associated With Oil

During the long learning process about finding oil fields, some areas were found by experience to contain natural gas but not oil. Eventually these terrains and, by analogy, other similar-appearing terrains, came to be carefully avoided during drilling for oil. With methane selling at 6 cents per thousand cubic feet, the discovery of it in modest amounts was usually no more economically rewarding than the discovery of salt water. Gradually, a set of circumstances was identified as probably indicating gas-only areas. Because these areas are now valuable as unexplored or lightly-explored regions, it is worth examining them in detail.

Higher Plants as Source Rocks

It has been established that most oil originates from the thermal cracking of organic material. The temperatures at depths of about 2.5 kilometers in sedimentary rocks are high enough to initiate the fragmentation of large organic molecules into the typical smaller molecules found in crude oil. If the sediment contains only a small amount of organic matter the resulting oil globules may be widely spaced and the oil-water interfacial tension will preclude the migration of that oil into reservoirs. However, in rocks containing the order of 10% or more organic matter, the oil created by the cracking process can form a continuous phase and migration from the organic-rich source rock into reservoir rocks can take place.

The type of organic material that most readily breaks down into oil is the remains of marine algae. In contrast, the higher plants that typically grow on land tend to produce smaller molecules during thermal degradation, usually leading directly to the formation of natural gas. This generalization is supported by the observation that coal beds do not tend to drip oil, even though we mine coal at every stage in its thermal maturation from peat through to anthracite. However, to the hazard of the coal miner, methane has a common association with coal as it is being mined.

It is not certain whether the predominance of higher plants in the organic matter is the cause of the paucity of oil in the great river deltas of the world. Of the many major deltas, only the Niger and Mississippi deltas are major oil producers. A re-evaluation of the great deltas as sources of methane will eventually be needed.

Gas in Tight Formations

The production of gas or oil from porous and permeable rocks follows the Darcy equation except in high velocity regions close to the well bore. The Darcy equation is simply a balance of a pressure gradient against viscous forces:

$$V = \frac{k \rho g h}{\mu L}$$

v = velocity
k = permeability
g = gravity
h = fluid head
ρ = density
L = length
μ = viscosity

The major variables in the ground are the permeability of the rock and the viscosity of the fluid. For even light crude oils the viscosity is about the same as water, of the order of one centipoise. In contrast, the viscosity of methane is about 100 micropoises, a viscosity 100 times smaller. Since the permeability is in the numerator of Darcy's equation and the viscosity is the denominator, there is a direct trade-off. Very much lower rock permeabilities can be tolerated in the production of natural gas.

The oilfield expression for a rock of low permeability is a "tight formation", and exploration programs for gas from low permeability formations are being undertaken in the Devonian shales of the Eastern U.S. and the Austin Chalk of Texas. Hydrofracturing techniques are increasingly important in improving the permeability close to the well bore so that low levels of permeability further out in the formation can still result in an economic flow of gas into the well.

Deeper than Five Kilometers

The thermal cracking process continues even after its initiation at 2.5 kilometers depth. Temperatures existing at about 5 kilometers depth are sufficient to break most of the long-chain hydrocarbons down to the short chains typical of mixed-gas or gas condensate reservoirs. The end result of the process is almost pure methane. Areas that are now buried beyond 5 kilometers depth, or which were once buried beyond 5 km and are now shallower, are an exceedingly important new source of methane. Because of its great importance, the discussion of these deeper and hotter parts of the sedimentary accumulations will be expanded in a later part of this chapter.

Overpressured Reservoirs

Normally, the pressure on the fluids found in sedimentary rocks is the pressure that would be caused by a column of water extending from the reservoir depth to the surface of the ground. In some areas, for instance in the deeper parts of the U.S. Gulf Coast and on Taiwan, higher fluid pressures are found which sometimes approach the pressure due to a column of rock extending from the depths to the surface. Experience both onshore and offshore on the Gulf Coast has shown that the overpressured section usually contains only gas condensate and gas reservoirs. It is quite possible that the abnormally high fluid pressure is due to the thermal dehydration of clay materials. If the typical dehydration temperatures of the local clays happens to fall at the transition temperature for oil reservoirs to gas-condensate reservoirs, then the observed association comes as no surprise.

Gas Clathrates

At low temperatures, water and a wide range of gases combine to form crystalline adducts. Two crystal structures exist, one with 6, and one with 5.67 to 17 molecules of water per gas molecule. In porous sediments in the Canadian Arctic and in Siberia, deposits of the methane clathrate exist. In addition, beneath the deep sea the sediments may contain methane clathrates. However, reliable estimates of the amounts of methane in clathrates do not exist, nor have schemes for recovering methane from clathrate deposits been brought into commercial practice.

These various terrains, whether characterized by plant-derived source rocks, tight formations, deep burial, overpressuring, or clathrate occurrence, are the grounds where we hope to support a new methane-based technology. There are additional sources of methane; other papers in this volume develop the possibilities for deriving commercial amounts of methane from these sources. For the next 10 years and probably longer, methane supplies obtained and developed with the drill bit and the rotary rig will dominate the available supply.

Thermal Maturation of Organic Matter to Methane

As an example of the kind of organic molecule that can be cracked thermally into fossil fuels, consider chlorophyll-A. The tetrapyrrole ring and the hydrocarbon "tail" of the molecule have rather similar counterparts in the porphyrin and phytene molecules in crude oil. This resemblance between living organic products and oil have long been known. Today, in addition, we can make some quantitative statements about the dependence of the hydrocarbon products on the thermal history of the surrounding sedimentary rocks.

In a qualitative way, the zone between 2.5 km and 5.0 km depth is referred to as the "oil window" (Figure 1). Organic-rich sedimentary rocks that have never been buried as deeply as 2.5 km are known as "immature" source rocks.

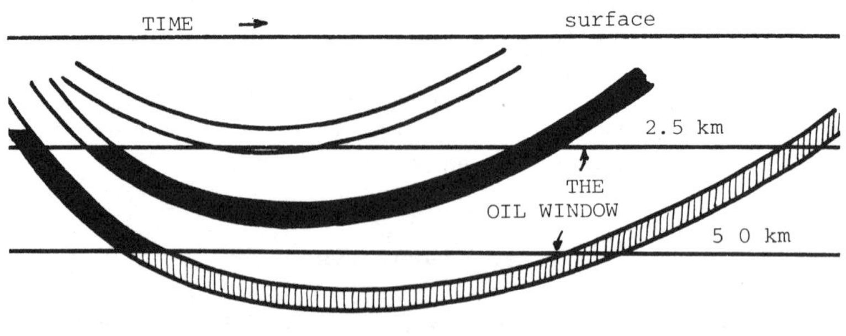

FIGURE 1

Those which have been buried, at some time in their history, beyond 2.5 km are known by petroleum geologists as "mature" source rocks. Burial beyond 5 km usually breaks the oil down to methane. The process is irreversible. Bringing oil-bearing rocks back shallower than 2.5 kilometers does not destroy the oil, nor does bringing deep methane-filled reservoirs back into the oil window recreate the oil.

Burial Histories

Although the concept of an oil window is a useful first approximation, there is an obvious dependence on time. If the burial is rapid, then higher temperatures are needed. Very slowly filling sedimentary basins might develop oil at lower temperatures. The dependence of the rate of a chemical reaction on temperature is given by the Arrhenius equation:

$$k = A \exp (\Delta H/RT)$$

k = rate of reaction
A = constant
ΔH = energy of activation
R = gas constant
T = temperature.

The Arrhenius equation only gives the momentary rate of the reaction - what we are interested in is the integrated effect of the entire time-temperature history of the sediment. In effect, we want to keep track of an "Arrhenius clock", a clock that runs faster in a manner consistent with the Arrhenius equation. (The concept of an Arrhenius clock was first stated in an Eastman Kodak advertisement about 20 years ago, apparently no further use was then made of the concept.)

We can consider starting with long hyrdocarbon chains and presume that the rate of thermal breakage will be proportional to the number of carbon-carbon bonds available to be broken:

$$dN/dt = -k N$$

N = number of carbon-carbon bonds
k = rate of reaction
t = time

This is the equation for a first-order reaction. We can then substitute the k in the Arrhenius equation in the first-order rate equation and have a first-order Arrhenius clock:

$$dN/dt = A \exp (\Delta H/RT) N$$

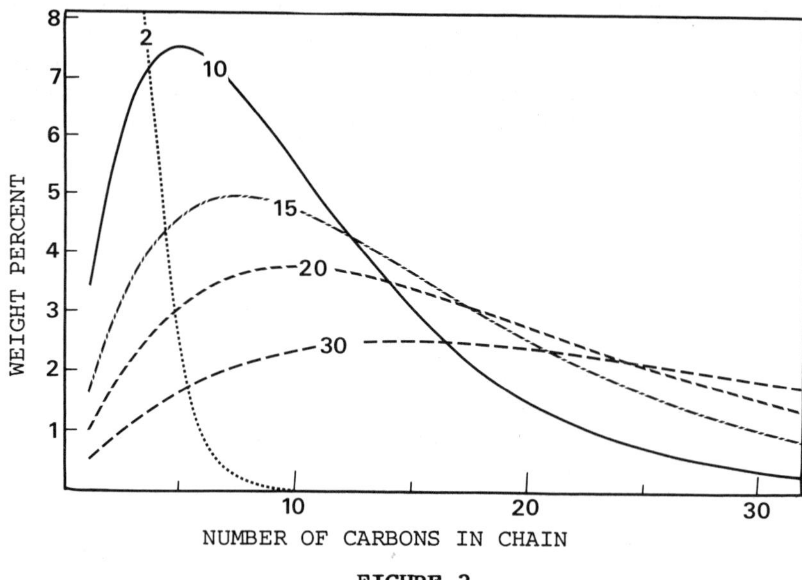

FIGURE 2

The numerical integration of this expression over the time-temperature history of the sediment gives the fraction of the original carbon-carbon bonds that remain unbroken at each stage of the history.

In order to get a picture of what the hydrocarbon chain length distribution would look like at each stage of the process, we can start with a very long chain and use some statistics to obtain the weight-averaged chain length. Figure 2 shows the chain length distributions.

Those characterized by mean chain lengths of 10 to 20 look remarkably like the chain length distributions in crude oils; the curve with a mean chain length of 2 closely resembles the composition of a "wet gas" containing primarily methane with lesser amounts of gas condensate. The almost-pure methane known as "dry gas" has a mean chain length of 1.1 or lower.

If the depth history were known, then the mean chain length and the expected discovery of oil or gas could be quantified. Until recently, it was virtually impossible to make any quantitatively useful statements about the burial history of sediments on continental platforms.

Fortunately, Tom Crough, and W. Jason Morgan at Princeton University have developed a useful hypothesis. They infer that in

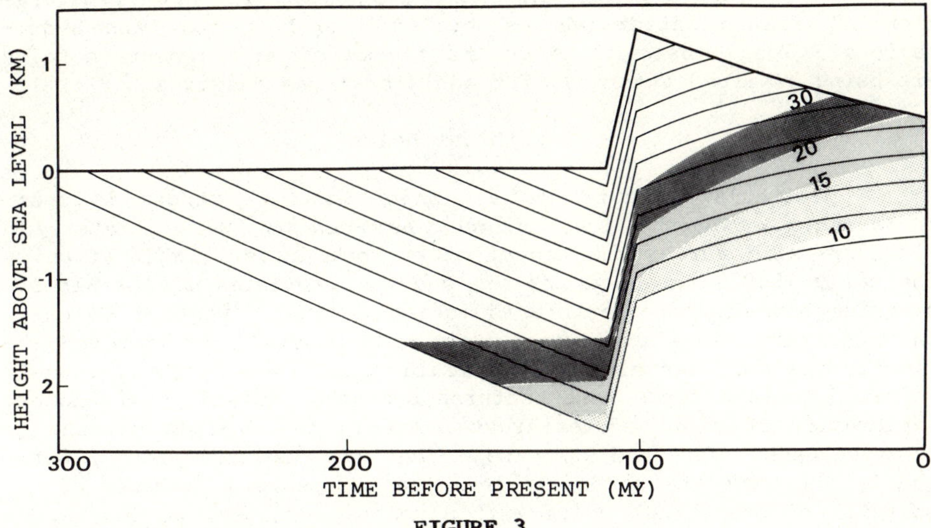

FIGURE 3

stable areas the upper mantle immediately beneath the continental crust slowly cools and slow subsidence and slow sedimentation will take place. Further, Morgan and Crough point out that passage of one of the heat sources in the mantle, such as in Hawaii, Yellowstone, or Iceland, will remove a portion of that cooled mantle layer beneath the continental crust. Even where the surface evidence of volcanic rocks is modest, a swath of upper mantle 1000 kilometers wide is typically removed, leading to a sudden uplift and erosion of surface rocks. Figure 3 shows a history, calculated by Silverman and Deffeyes, of sedimentation and uplift for an area in southern Kansas with mean chain length superimposed. Morgan is currently working on much more sophisticated computer programs for keeping track of several different Arrhenius clocks with different activation energies.

In addition to the uplifts and sinking of the continental interiors, there is another major thermal influence on sedimentary rocks. The cores of mountain ranges contain highly metamorphosed rocks and once-molten granites; strong evidence of intense thermal events within the ranges. Subtler evidence of thermal halos extending outwards from the cores of mountain belts was appreciated even before the turn of the century. As oil exploration proceeded from western Pennsylvania towards the Appalachians, oils were found to be lighter (shorter hydrocarbon chains) and eventually the oil was no longer found but natural gas was present. The correlation of this change with the change in the coal beds from sub-bituminous through to anthracite was the first evidence that thermal maturation of organic matter was associated with both coal and hydrocarbons. Today these areas on the flanks of major mountain belts are being examined intensely for additional gas reserves.

Overthrust Belts

A particular feature of many major mountain ranges is overthrusting, in which slices of rock as much as 100 kilometers in width are slid across the top of other rock units. Exploration in the early 1960's extended the oil and gas provinces of the Alberta prairies beneath the overthrusts in the mountains west of Calgary. In large part, this exploration was made possible by improved reflection seismic techniques that could "see" through the intensely deformed surface rocks to structures beneath. Now, 20 years later, exploration of similar overthrust belts in the U.S. is proving to be a success. (One can only hope that the methane-powered automobile can cross the border in less than 20 years.) Because of the position of overthrust belts within mountain ranges, we can expect more gas and lesser amounts of oil, to result from their exploration.

Additional Arrhenius Clocks

The hydrocarbon chains are by no means the only Arrhenius clocks within the sedimentary rocks. The rank of coals has already been mentioned, and quantitative measurement of light reflected from the vitrinite component of even microscopic amounts of coal is used as a quantitative measure of the integrated thermal history. Similarly, pollen grains change in color with deep burial.

One of the most useful thermal indicators, and one calibrated to the Arrhenius equation, is the colour of a class of almost-microscopic fossils known as conodonts. Anita Harris of the U.S. Geological Survey has produced thermal maps of the Appalachian and Great Basin regions based on conodonts.

A most important Arrhenius clock is the destruction of the porosity in the sedimentary rocks. Peter Weyl showed that the loss of porosity through pressure solution of grain-to-grain contacts depends on an integrated time-depth history. The existence of important natural gas reserves even below 20,000 feet establishes that porosity can persist to great depths. However, there is every reason to expect that the destruction of porosity will eventually limit the depth at which commercial amounts of gas can be developed. Some whole regions, like the Paleozoic sedimentary rocks of the Great Basin, may be unproductive because all of the porosity was destroyed when the section was more deeply buried than it is now.

Summary

Of the many possible sources of methane it is the author's present opinion that the greatest additional reserves will come from tight formations and from reservoirs that are now deep or were once deep.

TABLE 1. SOURCE OF METHANE

Depth	Sources
Surface	swamp gas, sewage, agricultural wastes, seaweed
Below 2.5 km	solution in oil, gas caps, tight formations
Below 5 km	deep reservoirs, overpressured reservoirs, solution in water.

These deeper and hotter parts of sedimentary basins and of the flanks of major mountain belts are volumes of rock only lightly explored during the search for oil. To be proved wrong because someone found even greater methane reserves in another category would only be a source of pleasure.

SUPPLIES OF METHANE IN THE UNITED STATES

W.J. Bowen

Chairman and C.E.O.

Transcontinental Gas Pipe Line Corporation

This chapter will summarize a major study completed by the American Gas Association's (AGA) Gas Supply Committee on the outlook for gas energy for the remainder of this century. The AGA worked on this study for two years; senior executives of many pipeline and distribution companies provided data and analyses and in addition, the Gas Research Institute and Institute of Gas Technology participated.

I believe the study is the most comprehensive and broadly based supply estimate ever developed by the gas industry. The study showed that the United States can increase its methane gas supply and, in doing so, can help reduce the amount of oil imports.

Figure 1 shows the U.S. energy consumption in quadrillion Btu's from 1950 through 1980. Coal, petroleum and natural gas quantities are shown. The category of "other" includes nuclear, hydroelectric, wood, solar, wind, etc. An average of recent estimates predicts that U.S. energy consumption by 2000 will be of the order of 100 quadrillion Btu's. This represents an increase of 24 quads and so, if oil imports are to be reduced, significant increases in domestic sources of energy will be required.

Today, the U.S. uses 76 quads of energy. The natural gas share of this 76 quads is approximately 25%. If that market share of 25% is maintained throughout this century, the 100 quad consumption level would imply that by the year 2000, 25 quads of energy will be required from natural gas or methane sources. However, the source mix will be quite different. Today in the United States about 95% of the gas sold comes from conventional wells in the lower 48 states. The rest of it is from supplemental sources of

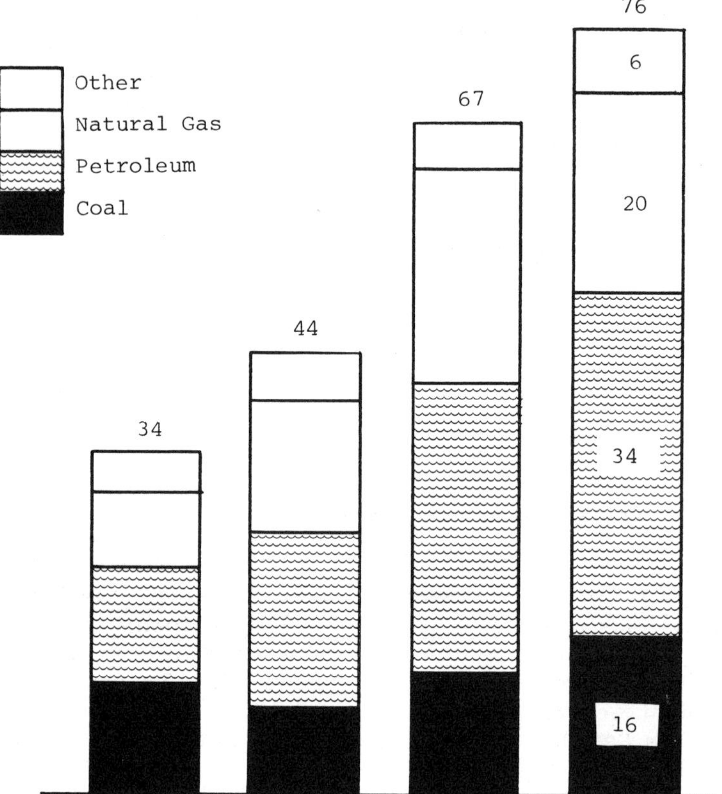

FIGURE 1. United States Energy Consumption
(Consumption - Quadrillion Btu)

which the largest would be importation of gas by pipeline from Canada. By 1990 we expect that conventional methane from the lower 48 states will be perhaps only 70%, the rest of it will be supplemental forms of methane, and by the year 2000 it could very well be at a 50:50 ratio.

Table 1 shows the various sources of natural gas or methane which were studied by the Gas Supply Committee. Each of the currently commercial sources are indicated. Included in this category are conventional production from the lower 48 contiguous states and supplemental sources from Alaska, Canada, Mexico, Western Tight Sands and Devonian shale gas; coal gasification; liquified natural gas (LNG) imports; synthetic natural gas (SNG) from liquid hydrocarbons and some gas from biomass sources.

TABLE 1. SOURCES OF NATURAL GAS

Currently Commercial		Not Yet Commerical
Lower 48	Western Tight Sands	Peat Gasification
Alaska	Devonian Shale	Oil Shale Gasification
Canada	LNG Imports	Geopressured Aquifiers
Mexico	SNG from Liquids	Biomass
Urban Waste	Coal Gasification	Coal Seams

Let us now consider the most important source of gas - the conventional production from wells from the lower 48 states. (Figure 2). Reserves have been declining because the production has been larger than additions to reserves each year since 1967. There was a peak in the production in 1972 but after that the decline in reserves was so great that the high production rate could not be maintained.

The decline in production caused gas prices to rise and drilling activity to increase. Figure 3 shows that gas well completions started rising very sharply in 1972 and, in 1973 a new

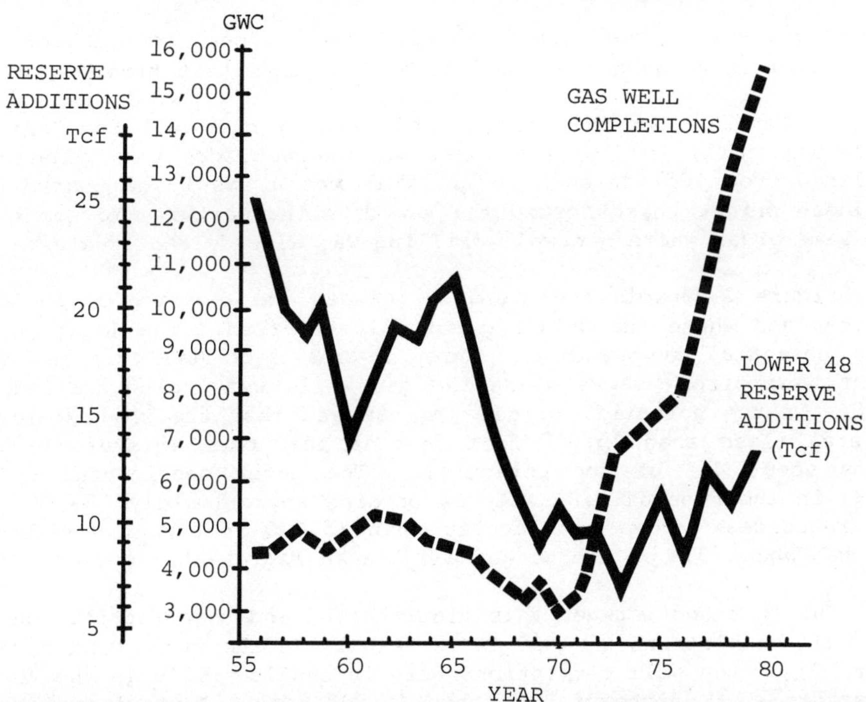

FIGURE 2. Lower 48 Reserve Additions and Gas Well Completions

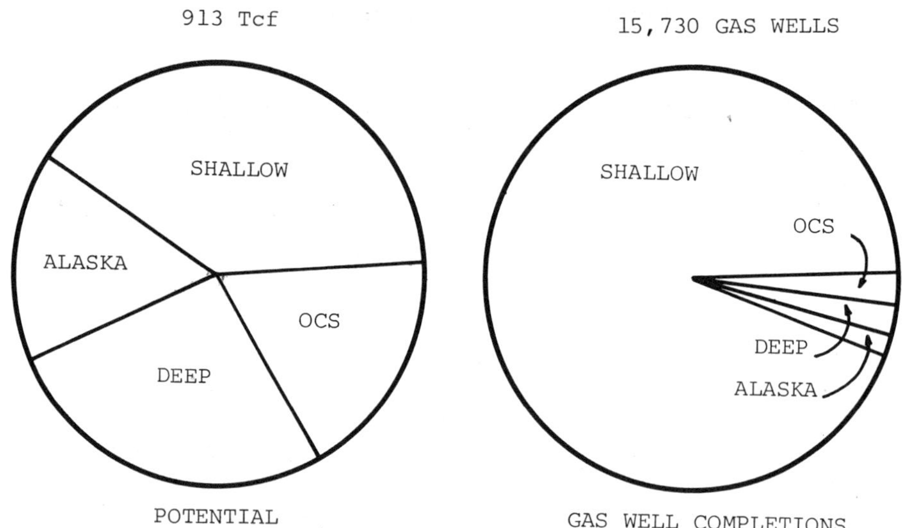

NOTE: "SHALLOW" AND "DEEP" REFER TO LOWER 48 STATES ONSHORE

FIGURE 3. Gas Potential and Gas Well Completions - 1980

record was set in the United States. New national records for gas well completions have continued to be set since that time.

In contrast to the very sharply rising gas well completions since the early 1970's, the lower 48 reserves additions generally declined from 1967 through 1977. This was a result of controlled wellhead prices which forced the new drilling activity to occur in the same areas where gas well drilling was already the greatest.

Figure 3 depicts the mismatch between the location of the resources and where the drilling actually occurred. The location of U.S. potential resources is shown in the left hand circle; the right hand circle shows where the gas wells were being drilled in 1980. On the potential circle you can see that the shallow lower 48 area (less than 15,000 feet deep in the lower 48 states) contains about 41% of the potential. The outer continental shelf (OCS) in the Federal U.S. waters contains approximately 17% of the gas resources. Deep gas (deeper than 15,000 feet), in the lower 48, has about 26% of the potential; Alaska represents another 16%.

The contrast between this distribution and the distribution of gas well completions in 1980 is striking. In the right hand circle over 90% of the well completions were in shallow wells in the lower 48 states - at depths of less than 15,000 feet. Approximately 3% were in the outer continental shelf area, about 4% were in the deep

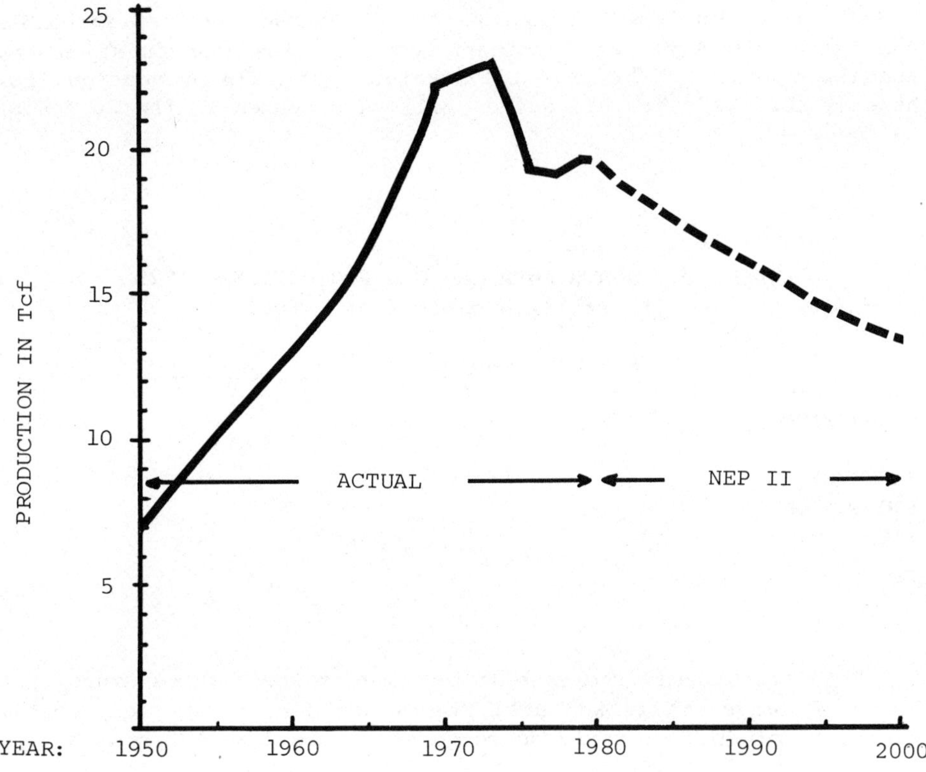

FIGURE 4. Lower 48 Natural Gas Production

areas, and a hardly discernible percentage was drilled in Alaska. This mismatch between where the potential is located and where the drilling has taken place is the principal reason the gas resources of the United States have not been more fully developed. The Natural Gas Policy Act of 1978 (NGPA) has started to change this situation but the overall drilling patterns change slowly. Probably the most dramatic change caused by the NGPA has been the increase in deep well drilling.

Figure 4 shows the actual and anticipated lower 48 production from 1950 to 2000. The 1980 and 2000 estimate is based on the analysis in the National Energy Plan II (called NEP II). This is the lower 48 production estimate which the gas industry accepts as a reasonable projection for what will occur in the last 20 years of this century. The actual production is shown in the left hand part of Figure 4 up through 1979. The dotted line shows the NEP II evaluation of the lower 48 production between 1980 and 2000. One important conclusion from Figure 4 is that the production from the lower 48 is expected to decrease. Obviously, supplemental sources of methane are needed to offset this decline.

Of these supplemental sources of gas, three that are available from the North American continent are pipeline gas from Alaska, Canada and Mexico. Table 2 illustrates that 1979 production from these areas. They are all quite small as compared to the 20 Tcf of the lower 48.

TABLE 2. NORTH AMERICAN GAS PRODUCTION - 1979 (In trillion cubic feet - Tcf)

	ALASKA	CANADA	MEXICO	LOWER 48	TOTAL
PRODUCTION	0.2	2.3	1.3	19.9	23.7
RESERVES TO PRODUCTION RATIO (R/p)	142.0	39.1	65.6	8.3	14.8

Production could increase in any one of these three areas by a factor of two or three and still there would be a very respectable reserve to production ratio as compared to the lower 48 states.

Canada has the capability of increasing exports to the U.S. very significantly from the current level of 1 Tcf per year to a level of 1 to 2 Tcf per year by the year 2000. Mexico could quite possibly increase gas exports to the same level of 2 Tcf per year. Both Canada and Mexico must determine what amount they consider surplus to their needs. The U.S. recognizes that Mexico may choose to maintain the same level which has been approved at the present time (300 MMcf) of gas per day which translates into 0.1 Tcf per year.

Clearly, however, the pipeline sources for North America are a secure, reliable and major source of natural gas for the future.

1978 world gas production, as shown in Table 3, excluding North America, was only 24.7 Tcf. Note that earlier we saw that North American production was 23.7 Tcf per year - consequently, North America is producing almost half of the gas produced in the world. The reserves to production ratio for the rest of the world is 90 to 1 and in the United States lower 48 it is 8.3. Therefore, there is a tremendous capability to increase the world production of gas. If world gas production is increased to a level approaching the intensity of production in the lower 48 states, there will be much more gas than the local markets can consume.

Under such circumstances, certain countries are in a position to export this gas in large quantities by either liquifying it and shipping it by LNG tankers, or by building pipelines which connect their production fields to the points of gas consumption. Of course, there is a very large capital investment to do this; this large capital investment limits the future of LNG in the world.

TABLE 3. WORLD GAS PRODUCTION
(excluding North America)

1978 PRODUCTION .... 24.7 Tcf

RESERVES TO PRODUCTION RATIO 90

PRODUCTION PROBLEMS: **Local Gas Markets Inadequate**
Must Export - LNG/Pipeline
Capital Investment

U.S. LNG import estimates are at the low end - 0.7 Tcf in 2000 which is just the gas available from U.S. projects which have already been approved. At the high end there is envisioned a vigorous world trade in LNG. It would be assumed that the U.S. would be a major participant in this trade.

An examination of world gas statistics makes it clear that the world use of gas is just beginning. Annual world gas production is now approximately 50 Tcf. World proved reserves are approximately 2573 Tcf, or enough for at least 50 years at current usage. Our current estimate of the total remaining recoverable gas (including proved reserves) is 9540 Tcf, or enough for about 190 years at current usage. As of year-end 1979 the total amount of gas used in history was approximately 1052 Tcf - just about 10% of the original total recoverable gas. So there is little doubt that current or increased production levels can be achieved well into the next century.

Many nations mindful of these statistics have started a dramatic movement toward gas. Recent studies have shown that:

- All OECD (Organization for Economic Cooperation and Development) nations which use gas, except for the U.S. increased their gas consumption from 1974 to 1978.

- Comecon nations for which data are available (U.S.S.R., Poland, Hungary, Czechoslovakia and East Germany have increased their consumption between 1970 and 1979.

In short, natural gas is the fuel of the future and abundant

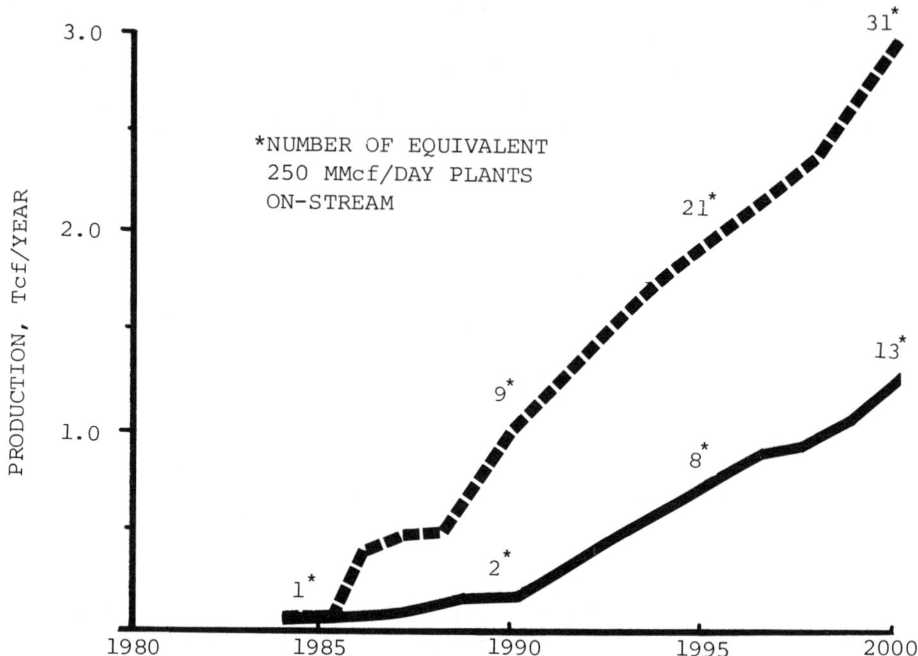

FIGURE 5. High-Btu Coal Gasification Industry Growth in U.S.A.

supplies of it can be developed if the world resources are brought into production. Demand for oil will be reduced and almost certainly there will be a decrease in the pressure for increasing world oil prices.

High Btu gas can be produced from the very large fossil fuel resources in the United States which represent between 28 and 45 thousand quadrillion Btu's. Coal is by far the most plentiful U.S. resource - representing about three-quarters of the remaining fossil fuel energy.

One major energy goal is to develop more environmentally acceptable and economically attractive ways to use this coal. One such use is coal gasification. Current technology permits the gasification of coal using the Lurgi process. The gas produced can either be produced as medium Btu gas or be converted to methane to produce high Btu, pipeline quality gas. The Lurgi process has been used in many parts of the world and is today in operational use in South Africa. If such plants were brought into operation in the U.S., large volumes of gas could be produced.

The first operational plant will be the Great Plains project in North Dakota. This plant, managed by a consortium of four gas pipeline companies, should produce pipeline quality gas (methane) in 1985. The U.S. Government loan guarantee, approved this summer, has provided a suitable financial environment to initiate construction work. This could lead to the construction of other plants, and by the year 2000 there could be as many as 30 high Btu plants producing a total of 2.8 Tcf per year. On the other hand, if support for such projects is not so enthusiastic, it could be predicted that as few as 13 plants might be built by 2000 with a production level of approximately 1.2 Tcf per yer. Although the buildup of the big production capability is 10 to 15 years in the future, these large quantities of domestically produced gas provide an attractive energy option.

Another extremely attractive domestic source of natural gas is the "tight formations". (Figure 6). The two dominant sources of tight formation gas are the Western Tight Gas Sands, largely in the Rocky Mountain states, and the Devonian Shale gas, largely in the Appalachian states. The volume of gas which is recoverable from these formations depends largely on the price of the gas and the technology available to produce it. Note that at a price of $6 per Mcf (in 1979 dollars) and with advanced technology, the production in 2000 from Western Tight Gas is three times greater than it is at a price of $3.12 per Mcf with the existing technology.

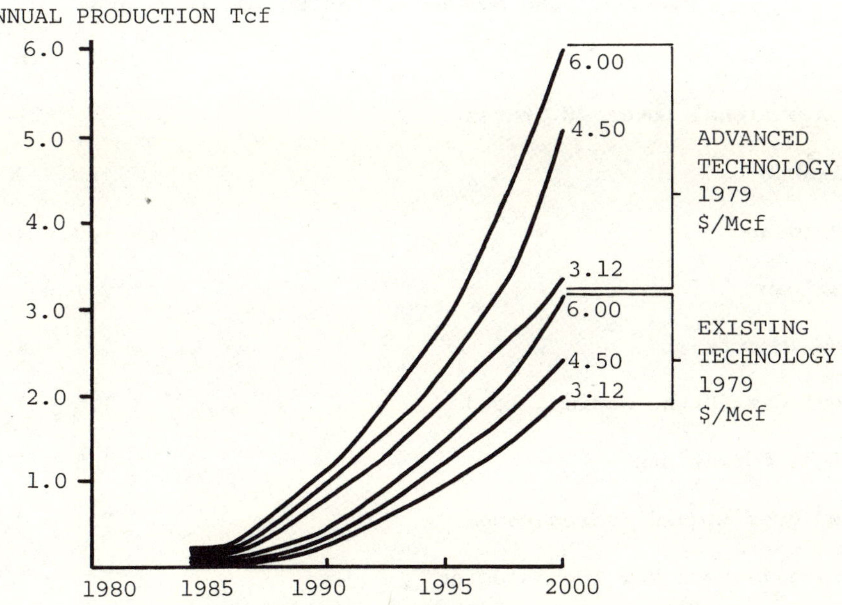

FIGURE 6. Annual Production Estimates from Western Tight Gas Sands

Note that very significant production levels from 2 to 6 Tcf per year are possible. These curves underscore the importance of a vigorous R & D program which will allow the advanced technology case to be realized. Current price incentives are in the upper portion of the range but even greater price incentives would be needed to achieve the full potential of this resource.

A similar situation exists for production from the Devonian Shale. Technology and price improvements pay big dividends in terms of the gas produced. However, in the case of Devonian Shales the levels of production do not exceed 1.5 Tcf.

Table 5 combines the range of estimates from each source of gas for production in the year 2000. If the assumptions for each of these estimates are compared, it is clear that the U.S. National policies which might lead to the high estimate from one source would not support the high estimate from another source. In other words, one cannot add all of the high estimates and get a meaningful result. Similarly, all of the low estimates cannot be added to produce a useful estimate.

TABLE 5. GAS SUPPLY ESTIMATES FOR YEAR 2000

	Tcf
Conventional Lower 48 States	12-14
Alaskan	1.5 - 3.0
Canadian	1.0 - 2.0
Mexican	0.1 - 2.0
LNG Imports	0.7 - 4.0
Coal Gas (High, medium, low)	1.5 - 3.5
Tight Formations	1.5 - 5.0
SNG From Liquid Hydrocarbons	0.1 - 0.5
Miscellaneous New Technologies	1.0 - 2.5

Source: A.G.A. Gas Supply Committee

To facilitate combining the various source estimates into a meaningful overall estimate, four U.S. national policy scenarios were developed. (Table 6) These policy scenarios are not meant to characterize our current policy, nor are they policy recommendations. They are policy options which serve to illustrate how the gas supplies from different sources change in response to U.S. federal policies.

TABLE 6. GAS SUPPLY SCENARIOS

SCENARIO	EMPHASIS	RESTRICTIONS
Self Sufficiency	Alaskan Coal Gas, R&D	Imports
North American Focus	Alaskan, Canadian Mexican, Coal Gas R&D	LNG
Moderate Imports	Canadian Mexican, LNG	Moderate R&D
World Conventional Gas	All Conventional Natural Gas Sources	SNG, Moderate R&D

The Self Sufficiency scenario assumes that federal policies encourage all domestic gas sources and discourage any imports. Naturally, Alaskan gas and coal gasification are key sources, as is the R & D necessary to develop new domestic gas sources.

The North American Focus scenario is based on a similar concept of emphasizing secure sources of gas supply except that Canadian and Mexican sources are encouraged. Any source of gas from North America is encouraged; imports of LNG from outside the continent are restricted in this scenario.

In contrast to the Self Sufficiency and the North American scenarios, the Moderate World Imports scenario would permit some limited amounts of increased LNG trade. However, restrictions would be placed on such imports so that they would not unduly affect our domestic gas supply.

The final scenario, World Conventional Gas Emphasis, envisions a world in which gas resources are being vigorously developed everywhere. As a result, large amounts of natural gas are available in many parts of the world for international trade. Competition would be so keen among nations exporting gas that the economics of LNG imports would be attractive. As one might expect, increased imports of this sort would undercut domestic R & D programs and so limit some of the domestic gas sources.

Table 7 arrays each of the four scenarios and the supplemental sources to show the overall results. Clearly, the greater the restrictions, as in Self Sufficiency, the less supply is available. It is interesting to observe that low values generally are between 23 and 27 Tcf; the high values are between 30 and 33 Tcf of gas. So the overall view of the scenarios does demonstrate a consistency of results. In the year 2000 U.S. gas supply should exceed 23 to 27 Tcf, but it will not be more than 30 to 33 Tcf.

TABLE 7. GAS SUPPLY SCENARIOS IN YEAR 2000 (TCF)

	SELF SUFFICIENCY	NORTH AMERICAN FOCUS	MODERATE WORLD IMPORTS	WORLD CONVENTIONAL GAS EMPHASIS
Lower-48	12-14	12-14	12-14	12-14
SNG From Liquid Hydrocarbons	0.3	0.3	0.3	0.1
Alaskan	3.0	3.0	1.5	3.0
Canadian	1.0	2.0	2.0	2.0
Mexican	0.1	2.0	2.0	2.0
LNG Imports	0.7	0.7	2.5	4.0
Coal Gas	3.5	3.5	1.5-2.5	1.5-2.5
Tight Formations	1.5-5.0	1.5-4.0	1.5-3.0	1.5-3.0
Misc New Technologies	1.0-2.5	1.0-2.5	1.0-2.5	1.0-2.5
Total	23.1-30.1	26.0-32.0	24.3-30.3	27.1-33.1

In terms of the original speculation about a total energy demand of 100 quads in the year 2000, if gas maintains its current market share of 25%, supplies of 25 Tcf are needed. That appears quite probable in all of the scenarios. If the market share for gas rises to above the 30% level, as it was in the late '60s and early '70s, then a gas supply of over 30 Tcf is needed. This is consistent with the upper limits of the four scenarios.

In summary, Table 8 estimates that between 23 and 33 Tcf of gas will be available in the U.S. in the year 2000. There is little doubt that this supply will be adequate to maintain the current 25% market share for natural gas. A larger market share for gas, of up to 30 Tcf, appears possible, if national needs dictate such a course of action.

TABLE 8. CONSEQUENCES OF SCENARIO ANALYSIS

Gas Supply in 2000	- 23-33 Tcf - About 25% of total consumption
Capital Investment Required	- Unprecedented - Gas provides the most energy per dollar of capital investment

Note: Federal/Regulatory Policies Are Major Variables in Supply Forecast

Unprecedented, high levels of capital investment will be required to support these supply projections. Despite the high levels of investment required, the capital efficiency of natural gas, as opposed to other energy sources, makes the selection of the "Natural Gas Option" highly likely.

The major uncertainties in these estimates are those associated with future U.S. federal and regulatory policy decisions. Technological uncertainties are also a factor but timely investment in R & D programs can reduce this to acceptable limits. One major effort to advance technologies critical to the natural gas industry has been the creation of the Gas Research Institute (GRI) in 1976. Some indication of its success and growth is that the total GRI annual budget already exceeds $100 million.

Despite this attractive supply situation there is a real concern that regulatory and legislative impediments to gas demand could effectively block the development of these gas supplies. Furthermore, these programs will be developed only if there is a strong demand for gas from the industrial sector and the gas industry remains financially healthy. The financial health of the gas industry in turn is tied to the maintenance of a strong industrial gas market, which provides a stable, non-temperature sensitive demand for gas service.

In addition to the regulatory and legislative adjustments to stabilize demand, the U.S. needs to develop a capability to activate key energy projects more quickly. One means of doing this would be to provide a special priority in terms of legislative, administrative and judicial implementation. Such "fast track" projects could become operational in a fraction of the time now required for complex energy programs. As well as providing more energy to the nations, "fast track" implementation would increase the economic viability of such projects.

Finally, this analysis concludes that pursuit by the U.S. of a vigorous program to develop a wide range of gas energy sources (i.e., "The Gas Option") is not only highly desirable but is also real and achievable. The U.S. does have the option of significantly increasing its gas supplies - reducing oil imports, reducing foreign dependence and reducing the flow of dollars overseas. This option is not a low probability long shot; it is a high probability future with adequate back-up in the wide variety of sources of gas supply.

EARTH OUTGASSING OF METHANE

Thomas Gold

Center for Radiophysics & Space Research

Cornell University

Summary

Earthquakes, volcanic eruptions, and a variety of other phenomena all point to a spasmodic release of gases generated at deep levels in the Earth. Methane (CH_4) appears to be a major component, as would be expected if the present meteorites are an indication of the source material that formed the Earth.

Oil deposits appear to have resulted from combinations of organic material and ascending gases from below; most of the natural gas that is now being discovered comes directly from the very deep (non-biogenic) sources, and has been arrested at accessible levels on the way up. The importance for the world energy situation is great, for one can expect to find very large quantities of gas in the sparsely explored but accessible region of 15-30,000 ft. depth.

Cosmochemistry and the Origin of the Volatiles

Carbon is the fourth most abundant element in the solar system after hydrogen, helium and oxygen. In the sun, in the outer planets and in their satellites, in comets and asteroids, carbon or carbon compounds are very much in evidence. The Earth and the other terrestrial planets, however, seem to have very little carbon in them. The deep rocks of the Earth seem to contain only small traces and carbon was also only a very minor constituent in the lunar rocks that were investigated. Yet on the surface of the Earth, and in the outermost layers, carbon is very abundant. Calcium carbonate and other carbonate rocks form a major part of the sediments that have accumulated in the course of geologic time. While most of the other constituents of sedimentary rocks appear to be derived from the grinding up and the weathering of the igneous rocks, the carbon content cannot be accounted for in this way: there is far too much of it in the sediments.

Did carbon then fall in late as the last addition to the forming Earth? Or did the carbon become concentrated on the surface through some carbon compound being gaseous or liquid and migrating upwards from the deeper layers?

The geologic evidence points clearly to an outgassing process bringing up carbon gradually over all or most of geologic time. The laying down of the carbonate sediments took place mostly from carbon dioxide in the atmosphere, dissolving in sea water and combining there with calcium oxide (and some other metal oxides) to form carbonates (limestone). At the present day this process occurs largely through the action of small organisms, but in earlier geologic epochs a very similar deposition took place without the intervention of biology. It seems that there has been a supply of carbon dioxide available for this process from the earliest epochs of the geologic record until the present. The total amount of carbon dioxide that would have been needed to create all the limestone is estimated to amount to about 50 kilograms per square centimeter of the surface area of the Earth, or about 50 atmospheres. Venus, the sister planet to the Earth, has about 90 atmospheres of CO_2. On the Earth, however, there never seems to have been a very large amount of CO_2 in the atmosphere at any one time, but rather carbon dioxide seems to have been supplied into the atmosphere gradually and put away into limestone more or less as fast as it was supplied.

Evidently, some process in the Earth is continuously exhaling carbon dioxide, or some other carbon compound that turns into carbon dioxide on the way up or in the atmosphere. What is the source of this material, what are the pathways by which it comes up, and what are the detailed physical and chemical processes that take place?

The meteorites and the comets, as well as the bodies in the outer solar system, provide many clues about the circumstances at the time of formation of the planets. A hydrogen-rich disc existed in which a temperature controlled condensation into grains took place, such that the more refractory materials condensed nearer the sun, and the volatile materials only condensed further out. The bulk of the material of the Earth represents a collection of the refractory substances and very little of the volatile ones.

Elemental carbon is a refractory substance, but little of the carbon seems to have been in that form. Among the meteorites there is just one type, called carbonaceous chondrites, that contains much carbon, and there it is mostly in the form of unoxidized carbon compounds, including heavy hydrocarbons. This is also the type of meteorite that contains significant amounts of water and other volatile substances. The reason for the low abundance of carbon in the refractory materials must be that in a hydrogen-rich environment, hydrogen compounds became the dominant carbon-bearing material, and those compounds are mostly fairly volatile. The distribution of carbon in the forming of planets therefore followed largely the distribution of the more volatile constituents.

The fact that in the carbonaceous chondrites only very heavy hydrocarbons are found is not necessarily meaningful. These meteorites are small objects that have been flying in their orbits in space for 4 1/2 billion years, and whatever range of hydrocarbons they possessed initially, it is only the heaviest ones that could be expected to have stayed behind. Among the comets, which are perhaps made of somewhat similar materials, the light hydrocarbons are in abundance, and CH_4, methane, is strongly in evidence in the atmospheres of the major planets and some of their satellites. In the nitrogen atmosphere of Titan for example, it has been stated that methane plays the same role that water does on the Earth: there are methane clouds shedding methane snow or rain; methane

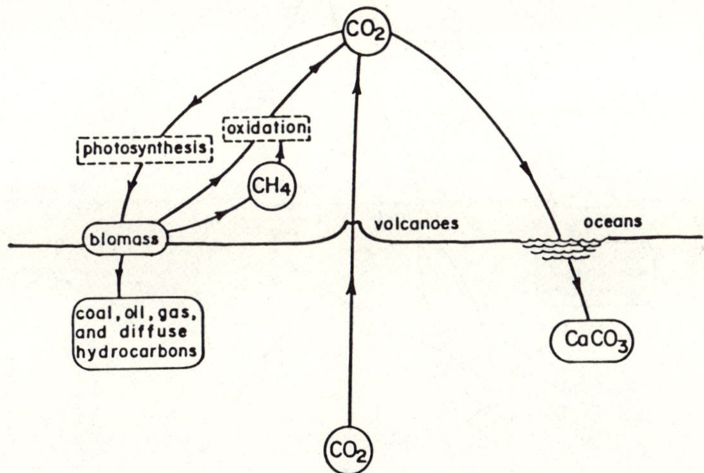

FIGURE 1. Terrestrial carbon budget. The conventional view holds that all coal, oil and gas is of biological origin, the product of photosynthetic reduction of atmospheric CO_2 into hydrocarbons which were buried before they could be fully oxidized. The initial source of the carbon is assumed to be deep-seated CO_2 which enters the atmosphere via volcanic emissions. The principal sink in the carbon cycle is precipitation through sea water into sedimentary carbonates, mainly limestone ($CaCO_3$). Almost all the atmospheric CH_4 is here assumed to be biogenic.

rivers flow on the surface and there are methane lakes and glaciers.

Like the carbon, the water on the Earth is also an outgassing product. The deep rocks are generally anhydrous, and would soak up water rather than push it out. So for both the carbon and the water one has to suppose that a volatile-rich material was supplied in the last stages of the Earth's formation, a material that was not characteristic of the bulk that had formed the main body. A material like that of carbonaceous chondrites would be the obvious one to suggest, both on the basis of it being the only meteoritic material that we know that would provide these volatiles, and also on the basis of it being the type of volatile-rich material that best fits the early solar-system chemistry.

A layer or a patchwork of this type of material must exist within the outer few hundred kilometers of the Earth. The great diamond-pipe eruptions and many other features support this conclusion.

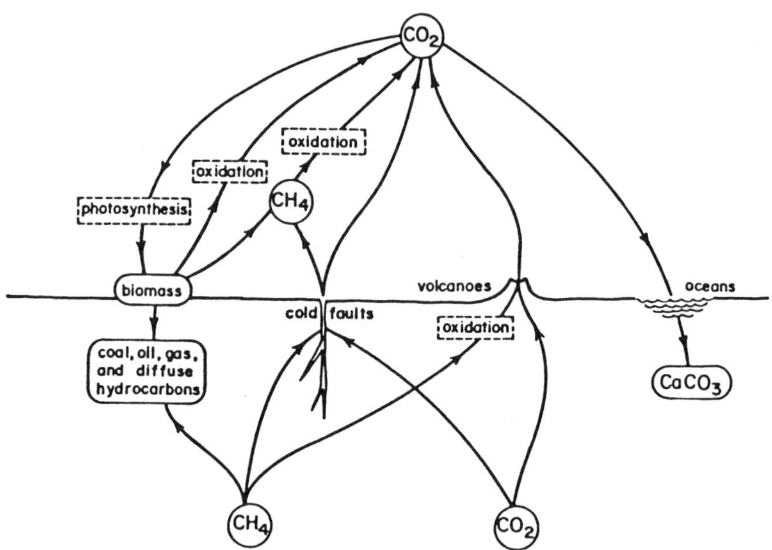

FIGURE 2. The duplex origin theory assumes, in addition to the photosynthetic reduction of CO_2, that there is a deep source of abiogenic methane contributing directly to the growth and maintenance of coal, oil and natural gas deposits. Some abiogenic methane is oxidized to CO_2 as it reaches the surface through hot volcanic lava at low pressure. But CH_4 can also enter directly into the atmosphere through cold faults, perhaps mainly during earthquakes, providing an abiogenic contribution to the atmosphere methane. Faults also provide a means for the seismic degassing of deep-seated CO_2 and other gases.

In what form would carbon be supplied from such materials? Under pressure and heat it would outgas carbon mainly in the form of methane. Other hydrocarbon compounds would exist also, but methane is the most stable one. Hydrogen gas and water would be other mobile products. In this picture one would thus have methane as the original carbon-bearing substance migrating towards the surface. Methane may also be formed from ascending hydrogen, together with immobile carbon compounds in the overlying layers.

This is quite a different outlook from that which has grown up in traditional geology. It has been believed generally that methane, as well as all other forms of unoxidized carbon that are found in the crust, have been derived exclusively from atmospheric CO_2, reduced in the biological process of photosynthesis. The question of the origin of the atmospheric CO_2 has had rather little attention, and most investigators seem to have been satisfied with the information that CO_2 is observed to be the main carbon-bearing gas that comes out of volcanoes. This observation, however, had nothing to say about the chemical form of the source material, since methane as well as any other hydrocarbon gases would all be largely oxidized into CO_2 in bubbling through liquid rock in the low pressure, high temperature circumstances of a volcano. The oxygen for this comes from metal oxides of the lavas. No case has been made that oxidized carbon was the main original source of all surface carbon, and from the viewpoint of cosmochemistry it seems rather unlikely.

Faults, Earthquakes and Outgassing

The gases coming up from deep may be seen with relatively

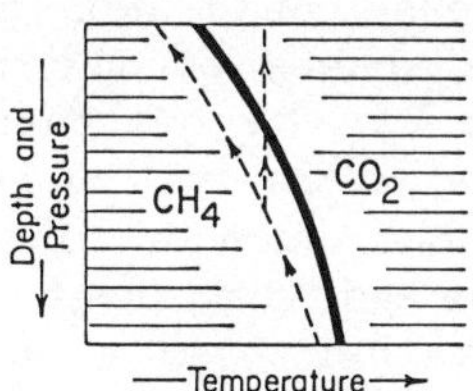

FIGURE 3. The pressure-temperature domain for the stability of methane and carbon dioxide in the presence of metal oxides of the rocks (diagrammatic only). At depth, where the pressure is high, methane survives at much higher temperatures than it would nearer the surface. Methane coming up along a volcanic pathway will thus be oxidized to CO_2 in the lavas at low pressure, but coming up along pathways that are cooler at shallow depths, it may survive as methane. The evidence that CO_2 is the main carbon-bearing gas that emerges from volcanoes is therefore no indication as to which of the two gases is the source material.

little modification at the cooler fault lines. Pathways through solid and not liquid rock can supply only a limited amount of oxygen from surfaces, and therefore oxidation would not destroy methane after some time. At deeper levels, where the pressure is very high, methane is considerably more stable both against oxidation and against dissociation and it appears, from the best modern data, that down to the temperature and pressure conditions at 300 km, and in the types of rock that are thought to dominate there, methane would be largely stable. Eruptions from depths of this order are thought to have been necessary to bring up diamonds, and the indications are that both methane and carbon dioxide existed there. Inclusions in the diamonds are known to contain both gases. The most likely origin of the very pure carbon forming the diamonds is then the partial dissociation of methane.

Major earthquakes provide circumstances where high pressure gases from great depths can suddenly gain access to the surface. Of course there is no one there, ready to sample and observe the composition, but some inferences can still be made. Gas eruptions have been reported from many or even most large earthquakes. Flames shooting out of the ground, on a small scale before a quake, as a precursory phenomenon, and on a massive scale during it, have been seen in many parts of the world, and in places where otherwise no petroliferous subsurface would have been suspected. Methane and hydrogen, as well as hydrogen sulfide and other gases, seem to be responsible, judging from observations surrounding the circumstances of some quakes. Methane and hydrogen are indeed the only gases that could come under serious consideration for providing the fuel for flames. Hydrogen is often present in volcanic eruptions together with small amounts of methane, and flames are observed there also.

The earthquake process could not occur without some fluid being present in the pore spaces of the rocks, at a pressure approximately equal to that of the overburden. It is only through this that a rapidly forming crack can be held open and allow sudden slippage to occur. Without such a fluid pressing it open, the friction is too high, due to the great compacting pressure, and only a slow distortion, rather than a sudden shock, would occur. This is true already for a depth of five kilometers, yet earthquakes are known with focal depths down to 700 kilometers. Methane and CO_2 in various proportions (and water vapor in the shallower cases) are the most likely pore fluids responsible. The ascent of these gases, in the processes of terrestrial outgassing, may well trigger earthquakes by weakening rocks that are under some stress. The various precursory phenomena that are now known can then be seen as arising from the escape to the surface of some of these gases, indicating merely that the rocks below have now been invaded by a fresh mass of gas, and that their maximum shear strength has now been greatly diminished; also that if a fracture were to take place it would be a sudden slippage and not a gradual distortion.

Derivation of Hydrocarbon Deposits

If then, as we believe, methane and its dissociation product hydrogen have been coming up towards the surface over most of geologic time, and if this has been responsible for supplying the bulk of the surface carbon, then of course the derivation of all hydrocarbon deposits in the ground has to be re-evaluated. It is then no longer clear that biology is the only, or even the principal agency for laying down hydrocarbon deposits. Much larger quantities of hydrocarbons have then streamed upwards from below than have ever been moved down again from surface biological activity. Nevertheless, a close and complex relationship between hydrocarbon deposits and biology is likely, simply because of the numerous interactions that hydrocarbons will have when they get into the biosphere. A duplex origin of many deposits, where the deep-source methane and hydrogen, as well as biology had a part, is then not at all improbable, and it is just such a mixed origin that fits best with many of the chemical details of petroleum and the regional distribution of the deposits.

Sir Robert Robinson, the distinguished British chemist, was aware of the chemical evidence for a mixed origin of petroleum. He wrote: (1)

> "Actually it cannot be too strongly emphasized that petroleum does not present the composition picture expected of modified biogenic products and all the arguments from the constituents of ancient oils fit equally well, or better, with the conception of a primordial hydrocarbon mixture to which bio-products have been added."

and also:

> "The absence of olefins ....strongly suggests that the oils have been subjected to powerful hydrogenating conditions."

Those investigators who attempt, in modern times, to account in detail for the occurrence of petroleum deposits from biological materials only, have encountered difficulties in particular in relation to the hydrogen content. Kerogen, the substance of biological origin in the sediments that is supposed to be the source of petroleum and methane, is very hydrogen-poor. A great amount of hydrogen enrichment would have to take place before the finely dispersed particles of tar-like consistency could be mobilized and made to flow to make a continuous petroleum reservoir. Tissot and Welte (2), in their book Petroleum Formation and Occurrence, write:

> "Processes of primary and secondary migration are so far poorly understood .... Therefore, it should be realized that the following discussion on petroleum migration is largely theoretical and should not be considered definitive."

Petroleum itself is fluid and almost fully hydrogenated, requiring two or three times as much hydrogen for each unit of carbon as the kerogen contained; and methane more still. It has been discussed that this hydrogen is derived from the cracking of yet more kerogen. Yet it is unlikely that the residue deposits exist, that would have to have supplied each oil or gas field. Such a hydrogenation process would, after all, be very inefficient, since the dispersed kerogen could not intercept and bind more than a small fraction of the diffusing hydrogen. It is very problematic how the great quantities of methane that are found could be created by such a process. If, on the other hand, the hydrogenation were to take place as a result of methane and hydrogen from deep sources streaming through, the problem would disappear. The methane that is found can then be attributed directly to this source; and where organic deposits have become enriched with hydrogen, there have been enormously greater quantities of gas available for this enrichment.

We suggest that major hydrocarbon deposits have formed in locations where the deep source gases have streamed through for long periods. This would then favor regional patterns corresponding to major underlying fault lines. The commercial deposits of hydrocarbons show such a relationship. Lines of earthquakes and the major volcanic lines are frequently paralleled by lines of hydrocarbon deposits. Mud volcanoes, which are the clearest manifestation of deep source methane escape, also follow such patterns. Hydrocarbon seeps or deposits exist in the vicinity of most volcanic zones: Alaska, Mexico, the Caribbean Islands and Venezuela are examples, as are the Aleutians (with seeps), Kamchatka and the Kuril Islands (where good measurements exist), Java and Sumatra, where the oilfields run the entire length of the volcanic belt and where they follow quite accurately the continuation of that same belt into the geologically different province of Burma. In so many instances it is clear that very large-scale underlying patterns are responsible for the distribution of the deposits, rather than the much more localized patchwork that would describe the distribution of organic sediments.

One way in which the accumulation of petroleum may take place is the following: along some deep fault lines methane and hydrogen seep up for long periods of time. Organic deposits of no particularly large quantity in the sediments trap some of this gas; methane is highly soluble, especially under pressure, in many organic oily substances, and hydrogen can react directly with unsaturated compounds. The fraction of these gases that is so arrested will then, in the course of time, add to the polymer structure of the organic oils and thus augment the amount as well as increasing the proportion of hydrogen. Temperature, pressure, catalytic agents in the rocks and microbial action may all play a part in these chemical transformations. The resulting liquids then become much thinner, and can be expected to flow in the interconnecting pore spaces of the sedimentary rocks. If this results in a continuous layer of pore spaces so filled, then all ascending gas will be arrested and the process becomes more efficient; the more petroleum has been generated the more effective it becomes in capturing the gas. Very large oilfields may then result, even though only very modest organic sediments existed in the region. In some instances the process may start without the advantage of an organic starter, and indeed there exists many instances of petroleum in the igneous rocks of the basement far from any sediments.

Although the origin of coal is widely believed to be completely understood, we think that it, too, has frequently (but not always) had a relationship to ascending methane. In circumstances where free hydrogen is allowed to escape rapidly, and where temperature, microbial and catalytic actions favor the dissociation of methane, carbon will be shed from the gaseous stream. This, then, can account for the great carbon enrichment that many fossils in coal have often suffered as well as the overall carbon enrichment of coal seams compared with the much wider distribution of insoluble substances that would normally co-exist with organic debris. Thick coal seams with a mineral content of only a few percent have always posed a problem to the purely organic origin theories. It must also be noted that many coal seams exist in which an occasional well-preserved fossil can be found, embedded in homogeneous coal. This alone makes it seem unlikely that it was other plants of the same kind and period, suffering the same treatment, that had resulted in the homogeneous coal deposit. In addition the fossils themselves show that they have been infused by some mobile form of carbon, so that the interior of every cell has become filled with carbon and the carbon content per unit volume has become many times higher than it was in the plant.

There are many areas where a strong regional relationship exists between gas, oil and coal deposits. In Indonesia regular sequences of oil and coal have been noted, vertically stacked above each other. Iran, Venezuela and Colombia, Alaska, the Appalachian mountains, Wyoming and large areas in Siberia all show a close regional relationship. In most of these cases the region is the more important factor and not the geologic epoch to which the deposit seems to belong. Thus the same region may show particularly rich deposits of gas, oil and coal, all vertically stacked above each other but spanning very long periods of geologic time. The San Juan Basin of New Mexico is a good example, where gas and oil in large amounts are found in deep Devonian sediments, more than 300 million years old, and shallow coal seams are above, belonging to Cretaceous (100 million years old) deposits. Why would it have happened so frequently that the same spot became favored with particularly rich organic sediments at totally different epochs when topography and all surface circumstances had changed completely? On the basis of these being areas of massive methane leakage, one can understand that the augmentation process would have been at work at all levels (but with results depending on the detailed local conditions).

In many areas there is a clear implication that the ground has been saturated with methane. It has been reported that all around the Arctic Ocean, every area of permafrost or of sea-floor deposits that has a pressure-temperature condition in which methane hydrates could be maintained, possesses such hydrates. Brines from the deep wells in Louisiana all seem to have a saturation value of methane content. Alkaline igneous rocks in Greenland, in the Kola Peninsula and in regions of Siberia show methane in occluded spaces. (In the oxygen surroundings of more acidic rocks methane would have oxidized when the rock was still liquid.)

Several of the greatest global rift systems have been seen, in recent times, to have water with a high methane content in lakes or oceans above them. The most celebrated case is that of Lake Kivu in the great rift of East Africa (3). Its deep waters have long been known to constitute the greatest methane anomaly: the methane content per unit volume is between one and ten thousand times greater than that of any other substantial body of water. Sediments are sparse, the entire region consists of young volcanic rocks and active volcanoes are on the shores of the lake. Is it a coincidence that this great methane content (2 trillion cubic feet) occurs just in one of the greatest rifts that are opening up at the present time? The northern extension of the rift system, the Red Sea, also has anomalously high methane concentrations in hot brines that emerge there.

The rift in the Pacific Ocean that is in the "East Pacific Rise" has been investigated by Dr. Harmon Craig of the Scripps Institute of Oceanography (4). It is a region far removed from any substantial sediments, and hot brines carrying a variety of gases emerge there. Craig has now found that at several places along this rift (and possibly along its entire length) methane comes up in substantial amounts. In Iceland, another place on a great rift, there are also some methane seeps that have come under investigation. Lake Baikai, in the great Siberian rift, has highly petroliferous regions on its shores in sediments of Cambrian age that would be regarded as quite unpromising if biological content were important.

The content of noble gases can give very important clues as to the derivation of any gas. Helium-3 in particular, being produced by terrestrial nuclear processes in much smaller abundance than it is found, must be largely a primordial component. Since there is no process that could have concentrated helium-3 on the Earth from the diffuse manner in which it was incorporated in the primordial material, it is an excellent indicator of the volume from which any particular gas sample must have accumulated. It is chemically inert and unlike helium-4, which results from the decay of uranium and thorium, it has no adequate terrestrial progenitors that may have been chemically concentrated. Therefore, if helium-3 emerges anywhere in unusually large amounts, it implies that it was "washed-out" from a large internal volume by another more abundant gas which carried it up through some vents. (By itself helium-3 is of such low abundance that gaseous streaming is out of the question, yet this is required to result in high abundance at vents). Craig has investigated the helium-3 content associated with the methane-rich region of the Pacific sources, of Iceland, and of Lake Kivu (4,5). In all three cases it was anomalously high, indicating that the gases had emerged from a large volume deep down. In the case of Lake Kivu the result was particularly dramatic: the water contained about three thousand times as much helium-3 as it would if it were merely surface runoff river water! The lake is thus at the same time the greatest anomaly both in its content of methane and of helium-3. It would be very hard to consider these two anomalies as unrelated.

Carbon Isotopes as Indicators of Origin

No discussion of the origin of hydrocarbon can be presented without referring to the carbon isotope information. Carbon on the Earth is approximately 99% carbon-12 and one percent carbon-13. This ratio, when investigated with high precision, is seen to have some small but significant variations. It is understood why the unoxidized carbon resulting from the reduction of atmospheric CO_2 in the photosynthesis of plants should be a little more deficient in the heavy carbon isotope, while the various deposits of oxidized carbon are generally slightly enriched. The processes that occur without the intervention of biology have effects of the same kind, but generally of smaller magnitude. The largest carbon isotope effects are observed in methane, and when isotopically very light methane is found (i.e. methane very deficient in carbon-13) it was generally thought that this proves its organic origin. This has become a cornerstone of the discussion, and therefore needs careful scrutiny.

It is probably true that in a single step process a larger

isotope selection effect can take place in biology than is likely in non-biologically-mediated reaction. But with methane the selection may be due to cumulative and not merely to single-step processes. By being a light molecule, the difference in the total mass, depending on the carbon isotope, is significant for many processes. The diffusion speed of the heavier molecule is three percent lower than that of the other, and this allows a cumulative selection to take place in the following manner.

If methane is steadily diffusing through the ground, towards the surface, there is no isotope selection: as many molecules of each type as are supplied below arrive at the surface, since in the steady state there cannot be an indefinite accumulation of any one type. But in such a diffusion column the mean density of the heavy type will build up to be greater by three percent, and this will then result in the same mean flow rates for the two, and no isotopic selection at the surface. If, however, some of the methane in the column is taken away by an oxidation process, then it is a different matter. The oxidation process may take the two types in the ratio of their density in the column, therefore favoring the heavy type by three percent. The larger the proportion of methane that is progressively lost in this manner on the way up, the larger will be the selection favoring the light isotope in the remaining methane. If the process could take place, then the observation that a sample of methane was isotopically very light would no longer prove that it was biogenic.

Now the evidence has been found, proving that such an isotope selection process is indeed frequently at work. In locations above known gas fields one has discovered in many instances a very particular type of carbonate in the ground: one that is isotopically much lighter than any of the ordinary carbonates (that are all derived from atmospheric CO_2) but a little heavier than the methane. This makes clear that it was derived from the oxidation of methane, as the association with gas fields already suggested, and it shows that it does indeed take away the heavier isotope preferentially. It is clear now that one cannot depend on the carbon isotope ratio to identify biogenic methane; the effect we discussed, by being progressive, probably goes even further in this isotope selection than any single-step biological process.

The Practical Importance

What is the practical importance of knowing the source of the carbon fuels? Will we find more as a result? Or is the search procedure that is normally used sufficient to find the fuels, however they came to be deposited?

It is true that a different theory of origin does not immediately suggest new methods of exploration. It is still necessary to find the traps in porous rock, and that has been the target all along. But there are some very important new elements. Firstly, one would search for gas irrespective of any biological content of the sub-surface. Regions that are now ignored because they are known to be free from organic deposits will come under consideration. Very ancient deposits are in that category. Secondly, one will have grounds for the expectation that the deeper levels will in general have more gas. Only a small fraction of the sediments have been probed at depths of 15,000 ft. or more, and there the success rate for gas has been very high. This can be attributed to good judgment, good luck, or a high incidence of gas at these levels. If it comes to be understood that the third alternative is the correct one, then one can look forward to a much more vigorous progress of deep exploration and drilling.

Criteria become important now that seemed irrelevant before. Deep underlying structures, indeed down to the base of the crust or even below that, will influence the patterns of seepage of gas. Deep and long-lived fault lines have an importance, the slopes of the deep layers or of the base of the crust itself may guide the upward moving gases and concentrate them at the high spots. Proximity to the faults that have resulted in volcanoes appear to be significant. Large-scale patterns in the deeper layers can be identified, and seem to be related to the seepage patterns of gases and liquids that have resulted in deposits of hydrocarbons and also of other minerals. Such patterns, and statistically very significant relations, tended to be ignored when, in the doctrine of the day, there seemed to be no rational explanation for them.

If the quantities of gas that come up are a major part of the surface carbon supply, then one is concerned with very much larger quantities than in the case of a biogenic origin, and one may expect more to have become trapped. Furthermore, if the supply has come from below and not from above, the occurrence of gas at or near the pressure of the overburden of rock is then a common feature at the deeper levels, and not an unlikely anomaly as it has sometimes been regarded. At the deeper levels there is a much better chance of containment, where the overburden pressure normally seals off any routes of escape unless they are forced open by an overpressure. At the same time the high-pressure gas can maintain an acceptable level of porosity in the regions it occupies. This pressure régime of sudden transitions to high-pressure domains with increased porosity is indeed what is found in many of the deeper wells, and could then be expected to be the rule rather than an anomaly.

Other than that, a new view of the origin of the deposits may also allow one to sharpen up some prospecting techniques. Regional trace element studies would seem useful, since the ascending methane (effectively a liquid at the high pressures at depth) may bring up in solution certain elements and concentrate them higher up. There have been indications that vanadium, nickel, mercury, and perhaps other elements show a regional relationship with hydrocarbon deposits. An understanding of such relationships would clearly spur a further use as a prospecting tool.

The hydrocarbon fuels are of immense importance to the present world. Efforts to understand their origin correctly should occupy a major position in applied research. The world could well afford a major effort, but it cannot afford accepting the prediction of a fuel shortage without even checking the basis of this prediction.

I am now confident that a vigorous program of scientific investigation, prospecting and deep drilling in the many unknown areas will soon show us that very much larger quantities of gas are available than are in the official estimates. A large-scale move to gas for stationary energy requirements is then indicated, so as to use the liquid fuels for transportation as long as possible. After that, conversion processes of methane to liquid fuels will be the next step.

REFERENCES

1. R. Robinson, "Duplex origin of petroleum," Nature 199, 113 (1963); and "The origins of petroleum," Nature 212, 1293 (1966)

2. B.P. Tissot and D.H. Welte, Petroleum Formation and Occurence, Springer-Verlag, New York (1978), pp 258-259.

3. K. Burke, "Dissolved gases in East African lakes," Nature 198, 568 (1963); and K. Tietze et al., "The genesis of the methane in Lake Kivu (Central Africa)", Geologischen Rundschau 69, 452 (1980).

4. H. Craig and J.E. Lupton, "Helium-3 and mantle volatiles in the ocean and the oceanic crust," in The Oceanic Liphosphere: Vol. 7, The Sea, John Wiley & Sons (1981), pp. 391-428.

5. Recent private communications with H. Craig.

6. T. Gold, "Terrestrial sources of carbon and earthquake outgassing." J. Petrol. Geol. 1 (3), 3 (1979).

7. T. Gold, and S. Soter "The Deep Earth gas hypothesis." Sci. Am. 242 (6), 154 (June 1980).

BIOMASS CONVERSION TO METHANE

Michael J. Antal, Jr.

Department of Mechanical & Aerospace Engineering
Princeton University
Princeton, New Jersey

Introduction

During the Arab oil embargo of 1973 a world-wide "energy-crisis" was proclaimed, but subsequent experience has revealed that it is really our technological addiction to the use of fluid fuels which is the underlying cause of energy malaise in the industrial world. Although great emphasis is presently being given to the conversion of solid fossil fuels (oil shale, coal, tar sands, and peat) into fluid fuels ("synfuels"), some scientists and policy makers have begun to ask the question: "Why not do the problem right and shift our reliance to renewable sources of fluid fuels rather than perpetuate our dependence on disappearing resources?" This presentation examines the possibility of answering their question affirmatively by using the renewable biomass resource to produce gaseous fuels.

The biomass resource is large. Recent estimates by the Office of Technology Assessment of the U.S. Congress (Figure 1) indicate that up to 17 Quads per year of biomass energy could be available by the year 2000. Thus at the national level biomass could be employed to generate about 85% of the U.S.A.'s natural gas demand.

The biomass resource can be even more significant at the state level. Figure I displays estimates of the biomass resource for New Jersey, which may represent the worst case analysis for the 50 states. New Jersey is the most heavily populated state in the U.S. with 953 people per square mile, as compared with a national average of 58 people per square mile. It has only 0.25 acres of commercial forest land and 0.1 acres of farmland per capita, in contrast to a national average of 2.4 acres and 5.1 acres respect-

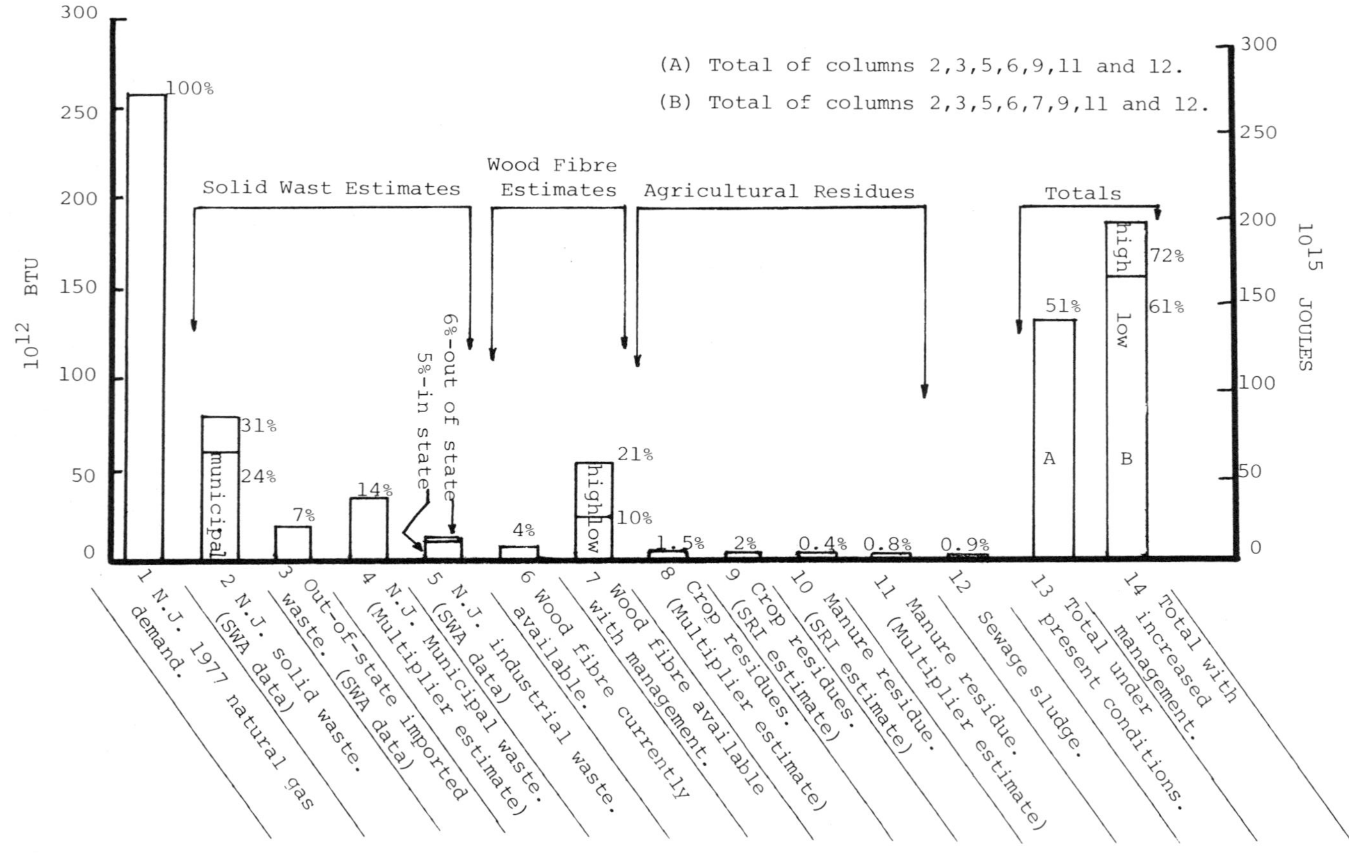

FIGURE 1. NEW JERSEY'S BIOMASS ENERGY POTENTIAL

ively. Nevertheless, there is sufficient biomass available in New Jersey to meet about 50% of the state's natural gas demand in the form of a medium-Btu gaseous fuel. This biomass primarily consists of solid wastes and wood wastes. Improved management of New Jersey's forestland could raise the biomass energy potential to about 70% of New Jersey's natural gas demand. Similar estimates for the state of Hawaii suggest that over 30% of the state's total energy demand could be met by biomass.

As a source of chemicals, biomass represents a truly staggering national resource. Table I shows that the U.S.A.'s readily available waste resource is sufficient to generate approximately twice its 1980 demand for ethylene using a rapid pyrolysis technology. Thus biomass can serve as a large scale source of chemicals and industrial feedstocks, as well as a source of fluid fuels.

TABLE 1. BIOMASS AS A NATIONAL SOURCE OF ETHYLENE

RESOURCE	ANNUAL PRODUCTION
Wood Manufacturing Residues	27×10^6 Dry Tons
Cereal Straw	161×10^6
Corn Stalks	142×10^6
Logging Residues	74×10^6
Total	405×10^6

Potential Ethylene Production @ 6% yield = 24×10^6 tons/yr

1980 PROJECTED DEMAND FOR ETHYLENE IN USA = 14×10^6 TONS/YR

These estimates all presume the use of efficient thermochemical processes to convert the biomass resource into a fluid fuel. Although inefficient biological processes (fermentation and anaerobic digestion) can be expected to play a role in specialized circumstances at the local level in the U.S.A., and an important role in some foreign nations, i.e. Brazil, they seem unlikely to contribute significantly to the industrial world's demand for large quantities of fluid fuels. Consequently, the remainder of this presentation focuses on the use of efficient thermochemical processes for biomass conversion.

Pyrolysis/Gasification Chemistry

Confusion often exists regarding the meaning of the word "pyrolysis". Strictly speaking, "pyrolysis" refers to the transformation of a complex molecule (like cellulose) into other, often simpler molecules (like ethylene) in an inert (non-oxidative) environment through the action of heat. Within the field of fuel conversion "pyrolysis" is often used synonymously with "incomplete combustion". Throughout this paper the strict meaning of the word "pyrolysis" will be assumed.

Pyrolysis is important as a biomass conversion method for two reasons: 1) it is the first step in the combustion or gasification of biomass fuels, and 2) it can be used to convert biomass into valuable chemicals and industrial feedstocks. Moreover, methane and synthesis gas (CO and H_2, which can be used to manufacture methane) are natural products of biomass pyrolysis.

Figure 2 illustrates the phenomenology of biomass pyrolysis. Upon heating, solid biomass materials begin to undergo pyrolysis at temperatures exceeding 300°C. This temperature is affected by the heating rate and the presence of catalysts. Some catalysts lower the temperature required for initiation of pyrolysis; whereas more rapid heating elevates the initiation temperature.

The solid phase pyrolysis mechanism is competitive: low temperatures produce primarily carbonaceous char, CO_2 and H_2O; whereas rapid heating of biomass gives high yields of "volatile matter" and little char. The volatile matter is thought to be chemically analogous to the depolymerization products of the complex polymers cellulose, hemicellulose and lignin. A recent article by Shafizadeh (3) gives an excellent discussion of solid phase cellulose pyrolysis chemistry.

Research at Princeton (4) has shown that little methane and no ethylene is produced during the solid phase pyrolysis of cellulose.

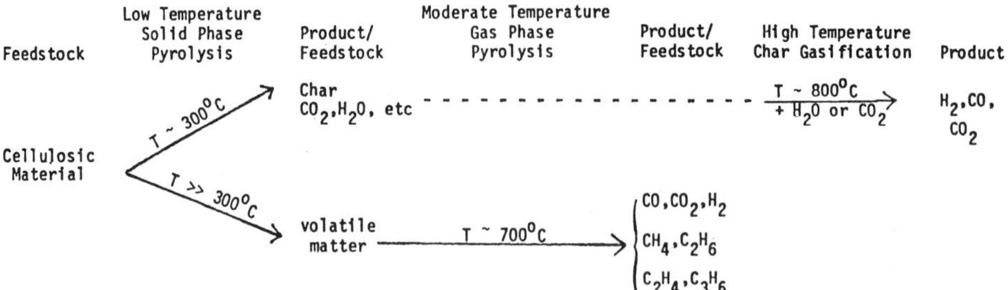

FIGURE 2

Levoglucosan (see Figure 3) is thought to be the primary, initial component of the volatile matter evolved from cellulose undergoing rapid higher temperature pyrolysis. In the gas phase levoglucosan undergoes further pyrolysis to produce permanent gases, including methane and ethylene. Figure 3 presents some speculations on the gas phase pyrolysis chemistry of levoglucosan. If these speculations are correct, it should be possible to obtain two molecules of ethylene per molecule of levoglucosan, or a 35% by weight yield of ethylene from cellulose.

In summary, the rapid heating of cellulosic materials produces primarily gaseous phase products which undergo further pyrolysis (in the gas phase) to yield methane, synthesis gas, ethylene, and (possibly) other valued chemicals. Virtually no knowledge exists of the gas phase chemistry, but there is some reason to believe that presently observed yields of methane and ethylene can be greatly enhanced.

Comparison of Biomass With Solid Fossil Fuels

The calorific value, moisture content, sulfur content and volatile matter content are widely accepted figures of merit of the quality of a solid fossil fuel. The following paragraphs offer a comparison of biomass to common solid fossil fuels based on available data.

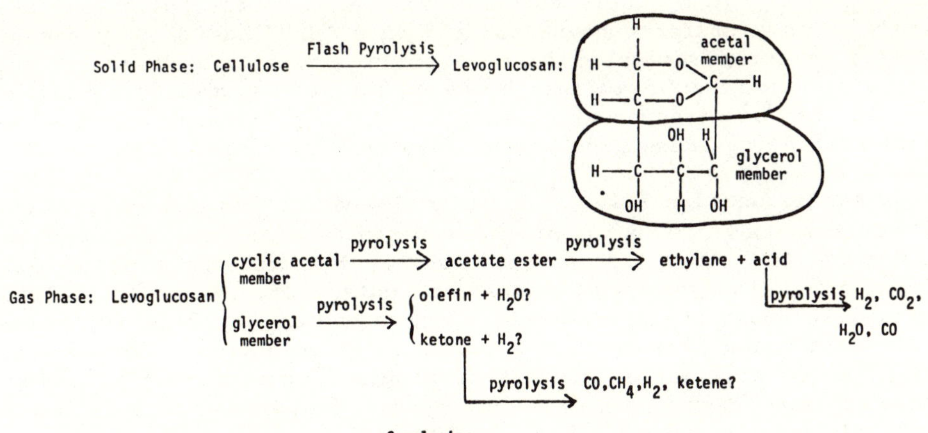

Conclusion

It should be possible to obtain 2 molecules of ethylene per molecule of levoglucosan, or a factor of four increase in experimentally observed yields.

FIGURE 3. Speculations on the Gas Phase Pyrolysis Chemistry

The calorific value of lignite is less than 19.3 kilojoules/gram (kj/g), sub-bituminous C coal ranges from 19.3 to 22.1 kj/g (8300 to 9500 Btu/lb), and sub-bituminous B coal ranges from 22.1 to 25.6 kj/g (9500 to 11000 Btu/lb). On this basis mixed municipal refuse with about 10.2 kj/g (4400 Btu/lb) is comparable to lignite, and dry woods (18.6 to 26.3 kj/g or 8000 to 11300 Btu/lb) are comparable to sub-bituminous C or B coals. The calorific value of agricultural residues is comparable to a good grade of lignite. Manure is also comparable to lignite. Based on calorific value peat is less valuable than wood, but compares favorably with refuse and manure. Of the fuels considered here, oil shale is the least desirable, having a calorific value of only 2.8 kj/g (1200 Btu/lb).

Moisture comprises typically 45% of the weight of lignite, 30% of sub-bituminous C, 24% of sub-bituminous B and 17% of sub-bituminous A coals. On this basis refuse compares favorably to a sub-bituminous A coal. Fresh wood contains between 25 and 50% moisture; whereas air-dried wood contains between 10 and 15%. Consequently, air-dried wood also compares favorably to a sub-bituminous A coal. Manure varies between 8 and 37% moisture, comparing favorably with sub-bituminous coals. Peat is typically 35% or more moisture and ranks with lignite as a relatively wet fuel. Oil shale has little or no moisture associated with it.

The sulfur content of U.S. coals varies between 0.2 and 7.0%. Low sulfur coal (less than 1% sulfur) is found primarily in the western U.S.A. Since the sulfur content of refuse and wood varies up to 0.1%, agricultural residues contain no sulfur, and manure contains 0.3% sulfur on a "dry" basis, refuse, wood, agricultural residues and manure cannot be compared to coal on the basis of sulfur content. These fuels outclass coal as a sulfur pollution-free fuel. With a sulfur content of 2.3% on a "dry" basis, only sewage sludge remotely resembles coal in pollutant value. Peat and oil shale are also low in sulfur; however both have significant amounts of nitrogen and are potential sources of NO_x pollution. Manure and agricultural residues also suffer from this problem.

As discussed earlier, the volatile matter content of a solid fuel is a good indicator of the ease with which the fuel can be converted to a liquid or gaseous fuel by thermochemical processes. On a dry basis, the volatile matter content of U.S. coals is less than 50%. Most organic materials are composed of 70 to 80% volatile matter and some contain more than 90% volatiles. Consequently, the volatile matter content of biomass suggests that it should be much more easily converted to synfuels than coal. Peat has a volatile matter content of about 70% (mcf basis), falling between coal and biomass in ease of volatilization. Oil shale is only about 10% volatile matter and is poorly suited for synfuel production.

In summary, the following is a ranking of the "intrinsic" suitability of common solid fuels for conversion to synthetic liquids and gaseous fuels by thermal processes:

1. biomass (most suitable for thermal conversion)
2. peat
3. lignite
4. coal
5. oil shale (least suitable for thermal conversion)

The above conclusions are in concord with a recent announcement by the Institute of Gas Technology (IGT) that peat is more easily gasified to synthetic natural gas than coal. Recognizing biomass to be the progenitor of peat, peat the progenitor of lignite, lignite the progenitor of sub-bituminous, bituminous, and semi-bituminous coals which are ultimately the progenitors of anthracite, it is not surprising to find biomass the most favored feedstock for synthetic fuel production. It is followed by peat, lignite, sub-bituminous, bituminous and semi-bituminous coals and finally anthracite. The only rationale for considering oil shale as a conversion feedstock is that conversion results primarily in the formation of liquid fuel.

Apart from the liquid form of the primary conversion products, oil shale is the least desirable solid fuel available in the U.S.A. today.

Generic Economics

As a solid fuel biomass has intrinsic value. Conversion of biomass to alternative fuel forms will be justified only to the extent that the conversion products have sufficient value to pay for both the feedstock and the cost of the conversion process. Thus the question of added value plays an important role in the economics of biomass conversion.

Figure 4 illustrates the yields and value of those products derived from pyrolysis of cellulose using a Princeton reactor. Higher yields of olefins have been reported in the literature; consequently, Figure 4 is thought to be a conservative estimate. Nevertheless, the added value is $63 per Mg, or three times the value of the feedstock. This $63 per Mg is available to pay for the cost of the conversion process.

The gaseous fuel products listed in Figure 4 have an energy content of 13.6 gigajoules; hence one ton of biomass is equivalent to 100 gallons of gasoline in the form of a gaseous fuel.

TABLE 1. GENERIC ECONOMICS OF BIOMASS PYROLYSIS

Input	Process	Yield	Product	Value
Quantity: 1 Mg	FLASH PYROLYSIS ——>>	56 kg	Ethylene	(1) $38
Feedstock: Cellulosic Material		11.4 Gj	Synthesis Gas	(2) 38
Energy Content: 17.4 Gj		0.6 Gj	Tars	(3) 1
Cost: $20.00		3.2 Gj	Char	(4) 6
	ADDED VALUE $63 PER Mg		Total	$83

* Composed of CO, H_2, CH_4, C_2H_6, C_3H_6 and CO_2
(1) marginal price = $0.68/kg
(2) price = $3.33/Gj
(3) price = $2.00/Gj
(4) price = $2.00/Gj

It is interesting to note that the value of ethylene comprises 46% of the total product value! Increasing the yield of the ethylene to the theoretical value of 35% would increase the total value of the products to $238 and the added value to $218 (a 245% increase). Clearly there is a strong economic incentive to produce more ethylene (or other high value products) from biomass feedstocks.

It is often argued that economies of scale necessarily make biomass uncompetitive, relative to coal, as a source of fluid fuels. This is because biomass cannot be transported long distances economically, and because biomass is rarely available at a single location in quantities exceeding a few hundred tons per day. This argument overlooks the fact that biomass is much more reactive than coal; consequently much higher throughputs result in lower capital costs and a less expensive product. In addition, the smaller scale of biomass gasifiers permits the use of mass production technologies using assembly lines to produce gasifier modules which can be assembled and erected much more cheaply that coal gasifiers. Recognition of these two facts underlies the development of oxygen-blown, pressurized down draft, biomass gasifiers at the Solar Energy Research Institute. The development of mass-produced, small scale biomass gasifiers is the most exciting and promising activity presently occurring in the field of renewable resources.

Biomass Energy Enhancement

For the past seven years research at Princeton has emphasized the use of concentrated solar radiation for biomass pyrolysis. Originally our motivation was to conserve the precious renewable fuel resource through the use of an external heat source. Since thermochemical conversion methods usually burn a fraction of the fuel resource to pyrolyze or gasify the remainder, the use of an external, renewable heat source in the conversion process has considerable value as a conservation technique. The West German program on nuclear coal gasification relies on this concept, using heat from a high temperature gas-cooled nuclear reactor to gasify coal.

Three years ago the President's Council on Environmental Quality commissioned a study on the use of solar heat for biomass conversion (7). A careful economic analysis in that study showed that simple conservation of the biomass resource using solar heat could not be economically justified. However, the study also showed that the use of concentrated solar radiation to effect flash pyrolysis of the biomass feedstock, enhancing volatile matter formation and reducing char formation, could be economically attractive. Since that time our research has emphasized the use of concentrated solar energy for the flash pyrolysis of biomass.

During the summers of 1979 and 1980, Princeton research using the Odeillo 1 MW_{th} solar furnace and the ACTF 340 kW_{th} solar furnace at the Georgia Institute of Technology (GIT) established the feasibility of using concentrated solar radiation to vaporize rapidly biomass materials in the transparent quartz reactor. Because the rates of vaporization were expected to be extremely rapid, the reactor design employed in both tests was based on a freely falling solids concept. Results of the Odeillo tests indicated that high yields of liquid syrups could be obtained from solar fired reactors; whereas the GIT tests provided high yields of a hydrocarbon-rich synthetic gas.

The ability of solar-fired chemical reactors to establish two characteristic temperatures within their working environment provided an explanation for the Odeillo results. Opaque biomass particles fed into the reactor were rapidly vaporized by the intense solar flux in a relatively cold, concurrently flowing steam environment. The hot vapors were quenched as they contacted the steam and were thus not subject to further pyrolysis in the gas phase. In the GIT reactor tests the flow of steam was countercurrent to the flow of biomass, causing the syrups to contact and condense upon the entering cold biomass particles before the syrups could escape the reactor. Thus the syrups were recycled into the reactor and subject to extensive pyrolysis, producing combustible gases.

Research at Princeton following these two tests has emphasized the production of syrups from biomass materials because these syrups are readily stored and easily transported. Thus the syrups can serve as an interface between the small scale, intermittent operation of a solar furnace and the large scale, continuous operation of a chemical refinery. The composition of the syrups suggests them to be a richer source of chemicals and fuels than crude oil.

Because the freely falling solids reactor did not provide sufficient control of residence time to insure complete pyrolysis of the biomass materials in all cases, recent research has emphasized the development of dilute phase, spouted fluidized bed reactors. These directly fired, quartz reactors provide sufficient residence times to insure complete pyrolysis. In addition, they can accommodate a wide range of particle sizes and are able to sustain a two temperature environment. Results of this research will be reported in forthcoming publications.

Conclusions

Conclusions 1 through 5 of this presentation are factual and do not reflect any bias of the author. Conclusion 6 may be biased, but evidence to date gives it credence.

1. Biomass is the most desirable solid fuel available in the world today. Its chemical properties make it preferred to coal, oil shale, peat or tar sands as a gasification feedstock.

2. The biomass resource (up to 17 Q/yr) is large enough to meet a large fraction of the U.S.A.'s natural gas demand.

3. The total U.S. ethylene demand could be easily met by the pyrolysis of available cellulosic wastes using available technology.

4. Basic and applied research on pyrolysis chemistry could increase ethylene yields from biomass by a factor of four.

5. Many other valuable chemicals (acetate esters, ketones, olefins, alcohols, alkynes, etc) could be obtained from biomass by pyrolytic methods.

6. Without a better understanding of pyrolysis chemistry, pyrolytic processes are unlikely to be a major contributor to the nation's energy supply.

Recommendations

The following recommendations were made by the author to the President's Council on Environmental Quality in 1978 (5), and again to the Office of Technology Assessment of the United States Congress in 1979 (6). As of this date, little progress has been made in the direction suggested: instead, greater emphasis in the U.S.A. is being given to oil shale and tar sands as energy resources. We can only hope that those few efforts devoted to the development of biomass succeed before our chronic addiction to fossil fuels becomes a terminal disease.

1. Basic and applied research on biomass pyrolysis as a source of chemicals, feedstocks and fluid fuels should be supported at a level comparable with coal gasification (at least $5 million per year).

2. Special emphasis should be given to novel conversion approaches which reflect the peculiar chemistry of biomass materials rather than the brute force conversion methods based on coal gasification chemistry which are presently in vogue.

3. Centers for research on biomass pyrolysis should be established at selected universities across the country.

REFERENCES

1. Bull, T.E. "Energy from Biological Processes", Office of Technology Assessment, Library of Congress, No. 80-600118, 1980.

2. Burwell, C.C. "Solar Biomass Energy: An Overview of U.S. Potential", Science 199, 1041, 1978.

3. Shafizadeh, F. "A Comprehensive Pyrolytic Process for Conversion of Wood to Sugar Derivatives and Fuels", Proceedings of the 1980 AS/ISES Annual Meeting, Phoenix, 1980.

4. Antal, M.J. "The Effects of Residence Time, Temperature and Pressure on the Steam Gasification of Biomass", in "Biomass as a Non Fossil Fuel Source", D.L. Klass (ed.), ACS Symposium Series 144, American Chemical Society, Washington, D.C. 1981.

5. Antal, M.J. "Biomass Energy Enhancement: A Report to the President's Council on Environmental Quality", 1978.

6. Antal, M.J. "Thermochemical Conversion of Biomass: The Scientific Aspects", a Report to the Office of Technology Assessment of the Congress of the United States.

METHANE FROM COAL CONVERSION

R.O. McElroy
Manager, Fuels Program
B.C. Research

Vancouver, B.C.

Introduction

Other chapters in this volume deal with potential future supplies of "natural" gas - i.e. methane recoverable from conventional and unconventional natural reservoirs. Regardless of how such supplies develop, there is an alternative route to methane through conversion of coal. This chapter presents data regarding the long-term economic availability of coal, brief descriptions of current technology for methane synthesis, and a speculative analysis of where large scale coal-to-methane plants are likely to be installed.

In the following discussion, "methane" refers to methane exclusively. In the literature, terms including high Btu gas, high calorific value gas, substitute natural gas (SNG) and methane are used interchangeably to describe fuel gases having methane as a major constituent and with a higher heating value of about 100 British thermal units (Btu)/cubic foot (approximately 21,800 kilojoules (kj)/cubic meter). It is reasonable to expect that in the future C_2 - C_4 hydrocarbon gases from coal conversion will be separated for use as chemical feedstocks.

Conventional Liquid and Gas Fuels

In this author's opinion, world economic availability of conventional petroleum and "natural" methane will continue to decline in real terms regardless of price trends, exploration efforts and conservation. The rate of decline is a subject for futurists and, in view of the current oil "glut", can reasonably be expected to be erratic.

Coal Availability

The recently completed World Coal Study (1,2) includes an analysis of availability of hydrocarbon resources generally and coal in particular. The first conclusion of this study is:

> "Coal is capable of supplying a high proportion of future energy needs. It now supplies more than 25% of the world's energy. Economically, recoverable reserves are very large - many times those of oil and gas - and capable of meeting increasing energy demands well into the future".

The actual rate of increase of coal available to feed utilities, steel mills, conversion plants, and other consumers is subject to numerous practical constraints including demand, capital commitment, and transport infrastructure. However, the "...economically recoverable reserves..." (documented and distinguished from "resources" which are of questionable economic value) are in the ground waiting for exploitation.

TABLE 1. World Coal Resources and Reserves, by Major Coal-Producing Countries

(millions of tonnes coal equivalent-mtce)

	Geological Resources	Technically and Economically Recoverable Reserves
Australia *	600,000	32,800
Canada *	323,036	4,242
People's Republic of China *	1,438,045	98,883
Federal Republic of Germany *	246,800	34,419
India *	81,019	12,427
Poland *	139,750	59,600
Republic of South Africa	72,000	43,000
United Kingdom *	190,000	45,000
United States *	2,570,398	166,950
Soviet Union	4,860,000	109,900
Other Countries	229,164	55,711
Total World	10,750,212	662,932

* WOCOL member

Source: World Energy Conference and WOCOL Country Reports

Note: 2000 mtce/year is the energy equivalent of the total OECD oil imports in 1981.

As shown in Table 1, coal reserves are concentrated in relatively few countries, mainly in the northern hemisphere. These concentrations reflect two factors. Geologically, the northern hemisphere has more large sedimentary basins, hence more favourable targets for coal exploration. Historically, industrialization occurred more rapidly in the northern temperate zones with consequent demand and technical capability for coal production being concentrated in zones of high energy use.

With respect to future exploration, countries with long histories of coal use (e.g. Germany, USSR, USA) are unlikely to develop new coal basins but highly likely to delineate additional reserves within known geological structures. In other areas (e.g. Central Africa, Indonesia) favourable results of reconnaissance studies suggest that coal resources at least are more widely distributed than the data of Table 1 suggest. In any event, geo-political concentration of coal reserves does not approach the concentration of (exportable) petroleum reserves in the OPEC states of the Middle East.

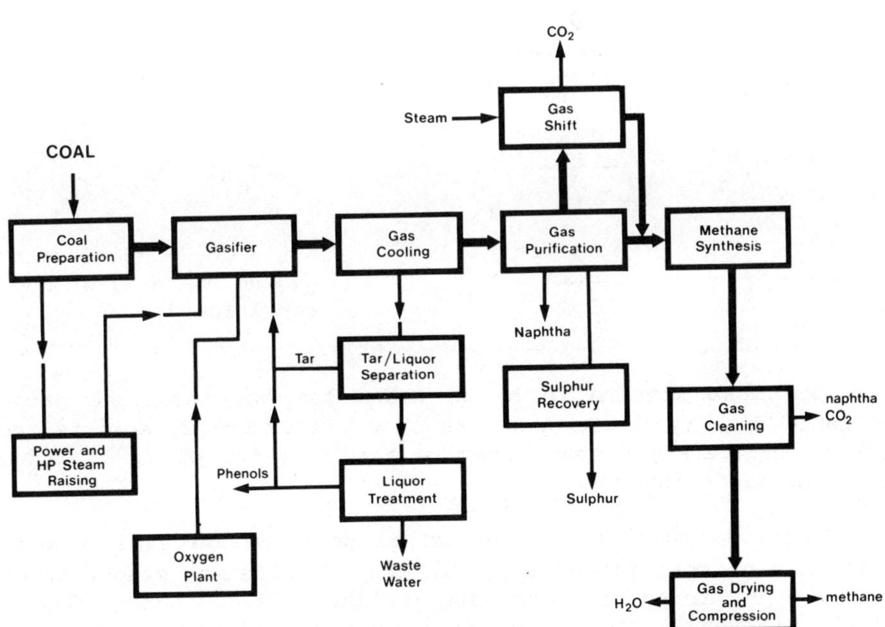

FIGURE 1. General Flowsheet for Coal Conversion by Gasification/Methane Synthesis

Coal Conversion - Technical Feasibility

In the broad sense, the term coal conversion applies to virtually all methods of utilization. Utilities convert coal to electric power, coke plants convert coal to special purpose solid fuels, and domestic users convert coal to heat.

Coal conversion to methane as a principal product by present technology involves the following sequence of operations, (see Figure 1):

Gasification: partial combustion of coal to produce a mixture of carbon monoxide and steam.

Gas Shift: catalyzed reaction of part of the carbon monoxide with steam to form carbon dioxide and hydrogen, normally followed by removal (stripping) of carbon dioxide.

Synthesis: catalyzed reaction of carbon monoxide and hydrogen to form methane and water.

In highly simplified form, these operations can be chemically described by:

$$\text{Coal} \xrightarrow[\text{steam}]{O_2} CO + H_2$$

$$CO + H_2O \xrightarrow[\text{P,T}]{\text{catalyst}} CO_2 + H_2$$

$$CO + H_2 \xrightarrow[\text{P,T}]{\text{catalyst}} CH_4 + CO_2 + H_2O \text{ (amounts depend on catalyst and conditions)}$$

Note that methane is by no means the only economic product available from synthesis gas. As shown in Figure 2, selected conditions and catalysts can generate a wide range of fuel products and chemical feedstocks.

Figure 1 indicates the principal unit operations involved in conversion of coal to methane. In the gasification stage, several commercially proven processes are available (e.g. Lurgi, Koppers-Totzek, Winkler). For methane synthesis, proprietary processes are also available. In fact, much of the research and development effort in synthesis catalysis has focused on ways to reduce methane yields in favour of higher hydrocarbons.

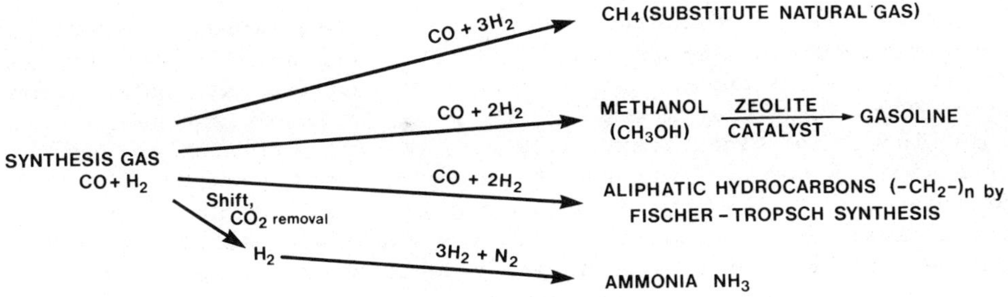

FIGURE 2. Alternate Uses for Synthesis Gas

In considering methane as a principal product or major co-product from synthesis gas, it is useful to review Lurgi fixed bed pressure gasification. This process, illustrated schematically in Figure 1, has been in use since the early 1930's and is - in various modifications - the most widely used coal gasification system. The gas product (Figure 3, after Sharman et al. (3)) contains approximately 8% methane by volume, but the methane accounts for approximately 24% of the calorific value due to its high molar heat of combustion. This initial methane product can be converted to carbon monoxide and hydrogen by reforming (further reaction with steam and oxygen) or partially avoided by gasification at higher temperatures. Both of these options involve significant losses in overall thermal efficiency and increased process complexity.

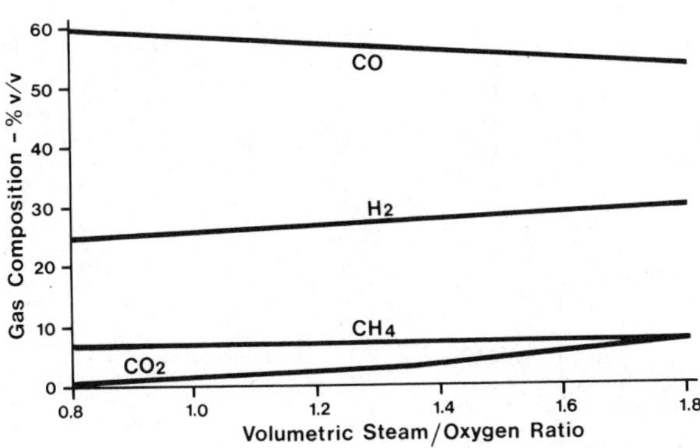

FIGURE 3. Composition of Dry Gas Product vs Steam: Oxygen Ratio (after Sharman et al., (3))

It is ironic that the only currently operational, commercial scale coal-to-hydrocarbon plants (SASOL I and II in South Africa) do not market methane. South Africa has no methane distribution system, so both plants crack methane to generate synthesis gas. This incurs a significant loss in thermal efficiency, but is economically valid in the specific circumstances. Actual operating efficiencies of SASOL type plants are in the 40 to 50% range but - for similar coals - overall efficiency could approach 60% if methane were marketable as a principal product or major co-product.

The existing SASOL plants and the proposed Great Plains SNG project in the USA both involve coal combustion to generate steam for gasification and process electrical power. Steam generation by nuclear power could increase the amount of methane (or other hydrocarbons) available per unit of coal. For higher cost coal, this could improve process economics as long clean burning hydrocarbons (e.g. methane) are suitably priced relative to electricity and coal. However, the multiple constraints retarding development of nuclear/electric power generation can be expected to have a similar effect on utilization of nuclear energy in hydrocarbon fuel processing.

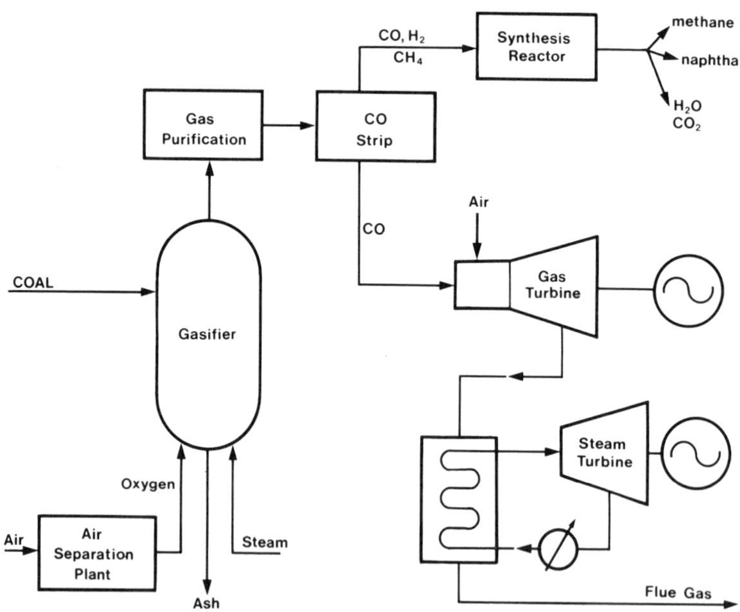

FIGURE 4. Gasification Synthesis with Combined Cycle Power Generation

With respect to multiple use of coal derived synthesis gas, there is an interesting option relating to combined cycle power generation. Synthesis of methane requires an H_2: CO ratio of up to 3 (Figure 2). This ratio can be adjusted by gas shift reaction or by stripping and separate utilization of CO. Conceptually, this could be accomplished in the type of operation illustrated in Figure 4. All of the methane (approximately 24% of fuel value of clean gas) plus all of the hydrogen (approximately 22% of gas fuel value) plus 13% of the carbon monoxide (approximately 2% of gas fuel value) are diverted to hydrocarbon synthesis, while CO is utilized in combined cycle (gas turbine/steam turbine) electric power generation.

Avoidance of the shift reactor stage improves efficiency, and the fuel gas disposition:

48%	-->	methane
52%	-->	electric power via combined cycle at up to 40% efficiency

could be attractive in suitably industrialized areas. At a somewhat smaller scale of operation, the separated carbon monoxide could also be readily utilized as an industrial fuel. Appropriately large fuel users include steel plants (direct reduction and/or blast furnace injection), petroleum recovery operations such as bitumen plants and secondary/tertiary recovery operations for conventional petroleum.

A full enumeration of developing coal conversion technologies is beyond the scope of this paper. However, in appropriate economic circumstances co-product methane could be available from processes such as:

Flash Hydropyrolysis:	low molecular weight aromatics, char, liquefiable petroleum gases (LPG), and methane
Catalytic Low Temperature Gasification:	methane, LPG, synthesis gas

Coal Conversion - Economic Feasibility

Cost of methane production from conventional wells in accessible, pipeline serviced areas is very low. However such supplies - even in Canada - are limited. Factors including tax policies, deregulation in the USA and rolled-in costs (Development plus transport) of new fields are pushing methane prices upward. In fact, incremental frontier production from such areas as the Alaskan north slope is expected to cost as much or more per delivered unit of energy than OPEC crude oil. If incremental frontier gas actually costs, for example, $7 (US)/10 Btu (equivalent to $42 (US)/barrel of oil), the competitive picture for methane from coal can be roughly sketched as follows:

sub-bituminous coal, calorific value 8000 Btu/lb delivered to mine-mouth plant	$10 (US)/ton or $0.62 (US)/$10^6$ Btu
conversion efficiency; (saleable products include methane, naphtha and phenols)	60%
product value (parity with frontier gas)	$7 (US)
fuel cost/10 Btu of product	$1 (US)

$$\frac{\text{Product value}}{\text{Raw material cost}} \simeq 7$$

The product value to raw material cost ratio of 7 leaves a large margin for capital, operating and management costs. For comparison purposes, a heavy or residual oil refinery operating in 1982 would have a maximum product value to raw material cost ratio of approximately 2, assuming high conversion to light, desulfurized products and minimal costs for supplementary fuel (e.g. methane for hydrogen generation). Such a refinery would have a number of unit operations (catalytic hydrotreating, fluid catalytic cracking, oxygen and hydrogen plants, gas treatment) and utility requirements similar to that of a coal-to-methane conversion plant.

It is reasonable to expect that the first commercial coal-to-methane plant will be high cost due to lack of experience relative to comparable ventures in the petroleum industry. Nevertheless, the large differential between raw material cost and product value (at parity with frontier gas) presents a significant opportunity for economic operation.

Potential Areas for Coal To Methane Conversion

The case described above is reasonable for coal deposits in natural gas pipeline serviced areas of the USA and Canada. However, given the large Canadian reserves of natural methane and the relatively low conventional reserves in the contiguous USA, an American site would clearly offer better potential. The Great Plains Project - if it goes ahead - will clearly define the economics of SNG production.

Applicable coal mining costs in Europe and the USSR are probably higher than for the case described above but in these areas the accessible markets are larger and supply situations - at least with respect to Western European dependence on OPEC oil - are potentially more critical.

Australia has suitable low-cost coal reserves and the basis of a continental methane distribution system. However, natural methane availability is high enough to allow large scale exports. South Africa has suitable coals and, through its SASOL plants, considerable expertise in the required technologies. However, South Africa has no methane distribution network and - in view of the methane recycle-to-extinction practice currently followed in SASOL plants - it seems unlikely that such a network will be developed unless substantial reserves of natural methane were to be found.

Coal to fuel gas plants are under construction at several sites in South America and Asia, but these are generally designed to service adjacent industrial complexes. In these cases, the methane synthesis stage is not economic compared to short distance pipeline transport of the clean fuel gas. Expansion of such plants (including methane synthesis facilities) would, in most cases, require very large investments in distribution systems and substantial changes in regional fuel use patterns. Given the relative shortage of investment capital in developing countries, such changes in fuel use patterns are considered unlikely.

Conclusions

It is apparent that technology is commercially available for conversion of coal to methane via the gasification/synthesis route.

World coal reserves are sufficient to provide feedstock for conversion at current or moderately increased methane consumption levels for some hundreds of years.

This combination of coal conversion technology and large reserve base provides an option for large scale methane utilization for a period much longer than the history of the current industrial era.

Economic application of coal-to-methane technology in the near term will require a number of favourable local factors including:

- a pre-existing distribution/marketing infrastructure for methane
- a large differential - probably at least a factor of four - in acceptable price of methane vs. cif. cost of

coal at conversion plant (prices and costs per unit of combustion heat)

- high incremental cost of methane from frontier, unconventional, or imported sources of "natural" methane
- availability of large amounts of investment capital.

Consideration of these requirements suggests that in the near future (perhaps to the end of this century) the USA is the most suitable area in terms of incremental methane prices, distribution/market parameters and capital availability. In the longer term, European operations - possibly at deep sea ports, utilizing imported coal - could become attractive as North Sea and USSR reserves of "natural" methane decline.

In South America, Asia, and Africa, lack of established distribution and marketing systems may preclude large scale methane synthesis, although coal conversion to clean fuel gas will continue to be an attractive option for large industrial fuel users.

For countries such as Canada and Australia with small populations and very large coal reserves, development of conversion systems will depend largely on export demand. The liquid methane export infrastructures developed in Australia and beginning in Canada may - in the long-term - favour methane production. However, production of synthetic liquid fuels will compete for capital in the export sector. Also, both countries have large current surpluses of deliverable natural methane which may inhibit commercial development of coal conversion.

REFERENCES

1. Coal-Bridge to the Future, Report of the World Coal Study. C.L. Wilson ed., Ballinger Publishing Co., Cambridge, Mass., U.S.A., 1980.

2. Future Coal Prospects, Country and Regional Assessments. R.P. Greene and M.J. Gallagher, ed., Ballinger Publishing Co., Cambridge, Mass., U.S.A., 1980.

3. The British Gas/Lurgi Slagging Gasifier, Springboard into Synfuels. 4th International Coal Utilization Exhibition and Conference, Houston, U.S.A., November 17-19, 1981.

TECHNOLOGICAL ADAPTIONS FOR ALTERNATIVE FUELS

CRUDE OIL CONSERVATION IN MOTOR VEHICLES

Enoch J. Durbin

Princeton University

Introduction

Natural gases are superb engine fuels and offer the ideal way to reduce our dependence on imported crude oil.

This is especially true since we need no new technology to implement a program of natural gas (NG) for motor vehicles. Engines have operated on natural and industrial gases since early in the last century and while there are limitations in the use of natural gases as fuels, these limitations can be diminished by new developments. At the same time, there are striking advantages in the use of NG as fuels and these advantages can be enhanced by new developments.

In this chapter and in the two succeeding chapters by my colleagues, G. Born of the University of British Columbia, and G. Karim, of the University of Calgary, we will develop the technological logic for these conclusions.

Perspectives on Engines and Fuels

We know the oil shortage is real and it can only get worse. The supply of crude oil is finite. We in the industrialized world have dispersed our population and industry, creating a strong dependence on individual transportation. Major increases in mass transportation can only make small reductions in our use of the personal car. A National Science Foundation study shows that a 17-fold increase in our mass transit investments would only reduce car use by 20%. We have built the easiest mass transit systems already and additional mass transit systems will serve fewer people and cost more per person. The personal car is here to stay and since it consumes about 40% of our oil, it has to be made more efficient.

The industrialized nations which have a major dependence on crude oil, unfortunately, do not produce all that crude oil. It is vital that alternate fuels be found and developed and that we devise means of conserving our use of crude oil if these industrialized nations are not to be held hostage by those nations able to export crude oil.

The purpose of this discussion, then, is to identify what is worth doing about engines and fuels from a crude oil conservation point of view. Clearly conservation ideas and fuel alternatives must satisfy our environmental concerns as well.

Let us first examine fuel economy in the automobile. Since the aim is reduction in use of crude oil we must consider the engine and refinery as one integrated unit. In discussing efficiency one must go from the raw source to the output work. Unfortunately, this is not often done because of the different economic interests of the auto manufacturer and refineries. The auto manufacturer attempts to achieve the least expensive auto product cost, which can satisfy the letter of fuel economy and pollution control regulations. Unfortunately these regulations do not consider the refinery at all, and have resulted in an enormous waste of crude oil, as we shall see.

The thermal efficiency of an engine is defined as the ratio of the mechanical work output divided by the heating value of the fuel used. The thermal efficiency increases as the compression ratio (CR) increases. By how much? We can predict the improvement theoretically, but it is far more illuminating to see this experimentally.

Table 1 shows the results of experiments to find the real effect on fuel consumption of changes in compression ratio, and fuel octane number (ON). Seventeen vastly different engines from 1.2 to 4.2 liter capacity were tested to find changes in km/liter as you change CR of the engine. The required ON was measured for each compression ratio and the resulting fuel economy was measured. Upper numbers are engine dynamometer results, lower numbers are steady state vehicle results. At the bottom of the figure we see that the average ON change required per unit CR change is 5.6. The average fuel consumption reduction is 7.6% per unit increase in CR and 1.3% per unit increase in ON. Raising the CR of the average car by 4.0, from the present level of about 8.0, to 12 would require an octane increase of about 20 and would reduce fuel consumption by about 25%.

TABLE 1. EFFECT OF CHANGE OF COMPRESSION RATIO AND OCTANE NUMBER ON FUEL ECONOMY

TEST BED RESULTS (STEADY SPEED)								
ENGINE							% INCREASE IN FUEL CONSUMPTION PER UNIT DROP IN	
CODE	CAPACITY (CC)	CR RANGE	CR SPAN	OCTANE REQUIREMENT RANGE	OCTANE SPAN	ON/ CR	CR	ON
A	1800	6.9/9.5	2.6	87/97°	10	3.8	-	-
B	1275	8.0/9.8	1.8	90/97	7	4.0	-	-
C	1275	7.9/9.4°	1.5	88/96°	8	5.3	6.0	1.1
D	1500	7.5/9.0	1.5	90/97	7	4.7	5.7	1.2
E	2000	8.0/9.3	1.3	90/97	7	5.6	7.3	1.3
F	2000	8.5/9.5	1.0	90/97	7	7.0	8.7	1.2
G	1600	7.5/8.7°	1.2	90/97	7	5.8	8.3	1.4
H	1300	7.8/8.7°	0.9	90/97	7	7.8	9.8	1.3
I	1496	7.5/9.6°	2.1	86/96°	10	4.8	3.4	0.7
J	1502	8.0/10.3°	2.3	90/96°	8	3.5	4.1	1.2
K	4236	7.6/10.0°	2.4	90/96°	8	3.3	5.1	1.5
L	1100	7.9/9.0°	1.1	90/97°	7	6.7	5.3	0.8
R	3442	7.7/8.4°	0.7	94.3/97°	2.7	3.9	12.0	3.1
S	4235	8.4/10.0°	1.6	97/97°	-	-	10.5	-
°Measured Values - All Others Are Nominal								
Vehicle Test Results (Steady Speed)								
M	1300	8.0/9.0	1.0	90/97	7	7.0	4.4	0.6
N	1300	7.6/8.5	0.9	90/97	7	7.8	13.3	1.7
O	1600	7.6/8.5	0.9	90/97	7	7.8	10.7	1.4
OVERALL EFFECT OF BOTH TABLES						5.6	7.6	1.3

Table 2 shows this in another way. CR is changed and steady state fuel consumption is measured. We see that changing CR from 7.5 to 9.6, about 2 CR, reduces fuel consumption by about 20% in steady state driving. The gallons used in this experiment are imperial gallons. Note the very high miles per gallon achieved. These are due to the use of steady state tests. There are no accelerations, no braking, and no dynamic fueling errors. Dynamic fueling errors will be discussed later.

The reduction in fuel economy at higher speeds is due to the increase in air drag. These experiments show that you can save considerable fuel by driving slower. The U.S. 55 mph limit is an important fuel conservation regulation.

Higher CR leads to more useful energy per pound of fuel, but the energy required to move the vehicle depends on the weight of the vehicle, air drag, and friction of the engine, drive train and tires.

That is all we have to work with. Much effort has gone into reducing energy required per pound of load carried. In this effort the focus is on reducing vehicle weight, reducing air drag, and lowering friction. Future gains in this area are probably minimal. We now have low air drag vehicles, and there are limits to weight reduction for safety reasons.

TABLE 2. MEASUREMENTS OF STEADY-STATE FUEL CONSUMPTION

COMPRESSION RATIO	STEADY SPEED FUEL CONSUMPTION (MPG) 30 MPH	50 MPH	70 MPH
9.6:1	55.0	41.0	30.0
8.8:1	52.0	40.5	27.5
8.2:1	46.7	37.5	26.5
7.5:1	45.0	37.0	25.0

The principal tool for increasing useful energy per pound of fuel is by increasing effective compression ratio. Higher CR requires higher ON fuel, but not always, as we shall see later. How do we get higher ON? By anything which makes fuel burn cooler or more slowly, by additives such as lead, by more sophisticated refining (which involves breaking long chain molecules into short chain molecules) and by reforming molecules to make them more stable (which makes them harder to ignite and therefore slows the flame speed).

It is also possible to operate a higher compression engine with lower octane fuels by cooling the combustible mixture by adding water at the intake (this cools the intake mixture and lowers the temperature of combustion), or by adding more air (burn leaner) which also lowers the combustion temperature.

Increasing the flame speed by turbulence, or by a larger ignition source, reduces the tendency to knock with a given fuel by reducing the time available for the unburned portion of the mixture to self-ignite.

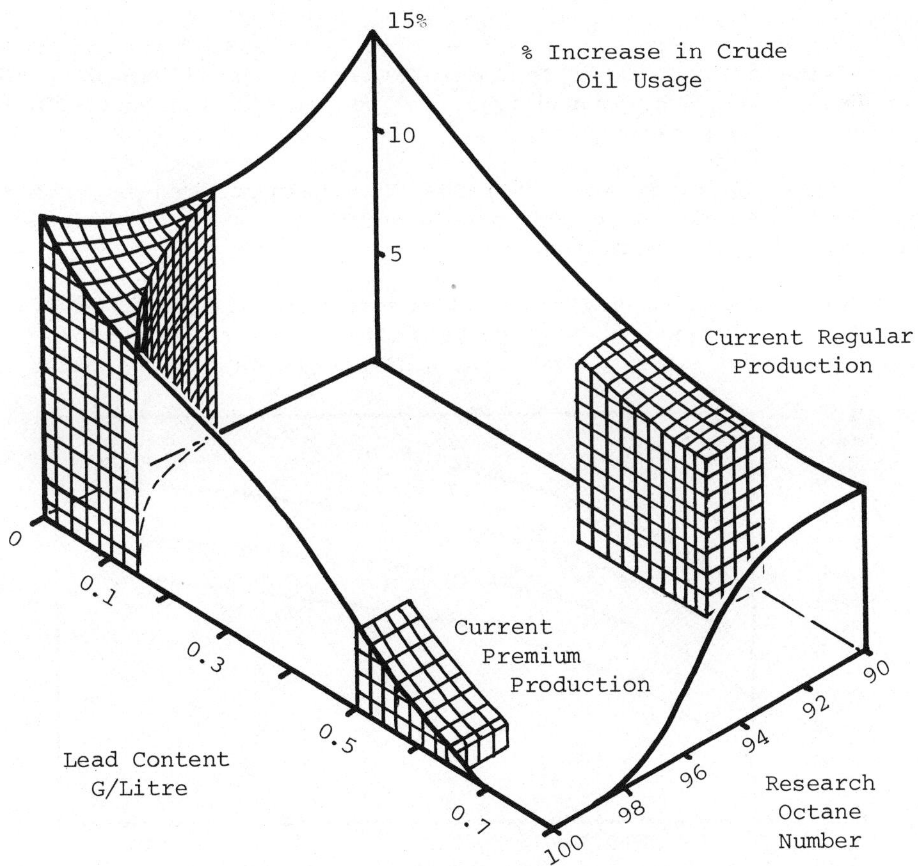

FIGURE 1. Relative Efficiency of Utilization of Crude Oil for Gasoline Production

Since the concern is crude oil conservation, one must examine factors which affect the amount of crude oil needed to make a liter of gasoline of a given ON.

Figure 1 shows refinery efficiency. Note the increase in crude used in making gasoline by raising ON without the use of lead. It takes about 8 more liters of crude per 100 liters of gasoline to make unleaded gasoline. At U.S. consumption rates of about 6 million barrels per day of gasoline, that is about 500,000 barrels per day of additional crude oil to make unleaded gasoline. We ought to use lead to raise ON and increase engine and refinery efficiency. Unfortunately our auto industry employs a platinum catalytic converter to clean up the exhaust pollutants, and platinum is easily poisoned by lead. Therefore there can be no lead when catalytic converters are used. With no lead, high octane fuel is difficult and expensive to make. The decision to employ an inexpensive catalytic converter to satisfy pollution laws forced the decision to require the elimination of lead in motor fuels. The consequential difficulty in making high ON unleaded fuel, in turn, forced the auto industry to lower CR from about 9/1 to 8/1. No lead and lower CR increased crude oil consumption by about 15%, or almost a million barrels per day.

There are other ways to raise ON. Methanol can be used in place of lead to raise ON, thus permitting higher CR while preventing catalytic poisoning.

In Figure 2 we see that 10% methanol is as useful as 0.4 g lead/liter in raising ON. Methanol has a heat of vaporization which is three times higher than gasoline. Its boiling point is

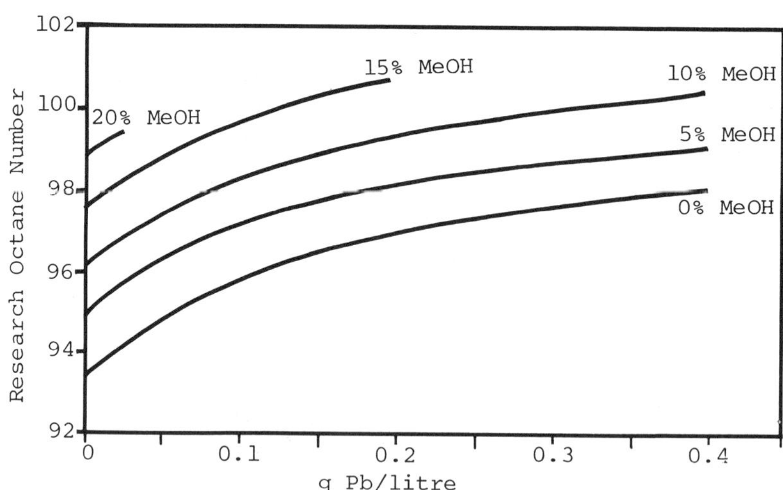

FIGURE 2. Methanol in Olefinic Gasoline RON Response with TEL

much higher than some of the more volatile gasoline components and, therefore, methanol has a cold start problem. The fuel temperature needs to be at least 80°C to ensure proper distribution of the fuel to the cylinders of the engine.

Methanol attacks zinc parts in the carburetor, fuel lines, and fuel tank. These problems are solvable, and methanol powered vehicles are used extensively, especially in Brazil.

We can draw conclusions from what has been stated so far. Changing the North American family car by raising CR by 4, from 8/1 to 12/1, and raising ON by about 20 could reduce our gasoline consumption by more than 25%. Further, if we also raised ON by lead or other non-energy intensive techniques, a saving of about 8% in energy consumed at the refinery would be possible. A total saving of 30 to 35%, or about 2.0×10^6 barrels per day, would result, which is about 1/3 of our crude oil import.

It is clearly advantageous to raise the compression ratio of our vehicle engines. What are the problems? Let us look at the operation of a four cylinder, 1.85 liter engine, on an engine dynamometer. The experiments show engine performance with a fixed amount of fuel per cylinder as we vary the air/fuel ratio and spark advance, the two principal engine control parameters.

Figure 3 shows the fuel consumed per unit output of work as we vary air/fuel ratio, and spark advance. λ , is the air/fuel ratio by weight. $\lambda = 1.0$ is the ratio where just enough O_2 has been provided to convert every fuel carbon atom and every hydrogen atom to CO_2 and H_2O if they were perfectly paired in the combustion process. This ratio is called stoichiometric.

The experiment is at constant speed. The fuel injected per cylinder is held constant for every data point of the figure. The air quantity and timing are varied. Note the great sensitivity to spark advance with lean mixtures. Lean operation requires accurate control of spark advance. The best fuel economy is achieved while operating lean. The fuel consumption is about 8% less than at a stoichiometric ratio. Lean mixtures burn more slowly; hence a greater spark advance is required and more importantly, as we shall see later, lean mixtures permit us to operate at higher compression ratios without raising ON.

In this same experiment we can look at pollutants. Carbon monoxide, CO, is the most deadly of the pollutants as it is taken up by the bloodstream, binds to the hemoglobin and reduces the O_2 distribution to our cells. It is severely regulated, is minimized with lean mixtures and is independent of spark advance. CO depends only on the availability of enough O_2 to make the CO into CO_2.

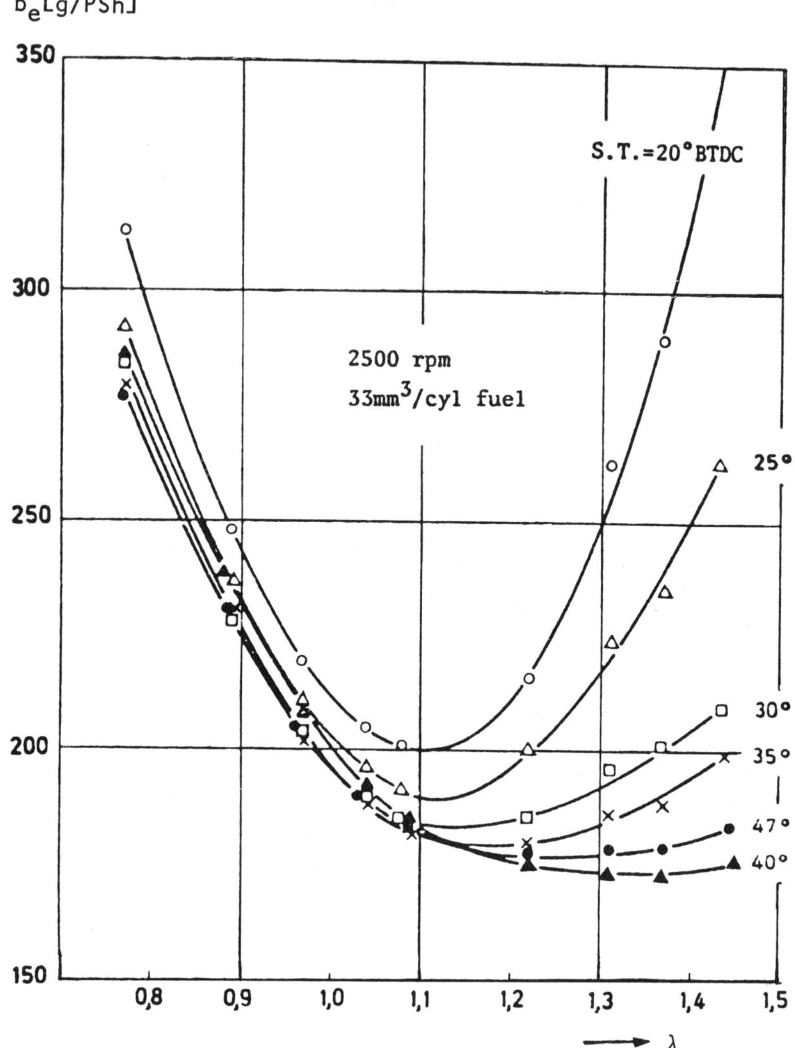

FIGURE 3. Variation of Fuel Consumption with Spark Timing and Air/Fuel Ratio

Hydrocarbons are the chief creator of smog in the atmosphere. Figure 4 is from the same experiment, and shows the variation of hydrocarbons in the exhaust gases as we vary air fuel ratio and spark advance.

Note that the hydrocarbons are minimized with lean mixtures. They are very sensitive to spark advance and are minimized with a retarded spark. Delayed completion of combustion makes hot exhaust gases, so that the hydrocarbons can continue to burn in the engine exhaust pipe. Hydrocarbons come mainly from quenching of the flame when it reaches the cold walls of the cylinders or the exhaust gas remaining from the previous firing of the cylinder.

NO_x turns into nitric acid in the atmosphere. It is also an important ingredient of smog, and seems to help create ozone which is highly oxidizing and attacks many materials.

Oxides of nitrogen, NO_x, are dependent on peak temperatures of combustion and how long they exist. High levels of nitrogen oxides are formed when temperatures are high for a long time. The formation of NO_x in the internal combustion engine is therefore a non-equilibrium process. Expansion of the combustion chamber by piston motion stops the formation process.

Figure 5 shows that NO_x is minimized by reducing combustion temperatures in lean mixtures by air dilution, and in rich mixtures with fuel dilution. The former is cheaper. The high spark advance required for good fuel economy also raises NO_x.

We have seen the value of lean operation, low pollution, and high fuel economy. Where do we operate? The decision to employ the catalytic converter as a clean-up device affected this decision as well. Figure 6 shows the performance of a 3-way catalytic converter. To operate the converter efficiently, we must run the engine at stoichiometric ratio. We have shown that we would rather operate lean but we are prevented from doing so.

What Kind of Engine Should We Be Building?

Figure 7 shows the theoretical thermal efficiency of the internal combustion engine as a function of λ and CR. The upper curve is the theoretical air cycle where heat is added at constant volume, or where combustion is instantaneous.

We see that for good economy we should operate at high CR and lean mixtures. The shaded area shows where we operate with current spark ignition engines. As the throttle is partially closed we operate at still lower effective compression ratios and are less efficient. This is due to air pumping losses which increase with throttle closure.

The upper part of the figure shows diesels operate at high compression ratio and with no throttling losses. The major difference in fuel economy between diesel and spark ignited engines occurs at part throttle conditions, where the unthrottled diesel does not have these high pumping losses. At wide open throttle conditions the overall efficiencies of both engines become more alike.

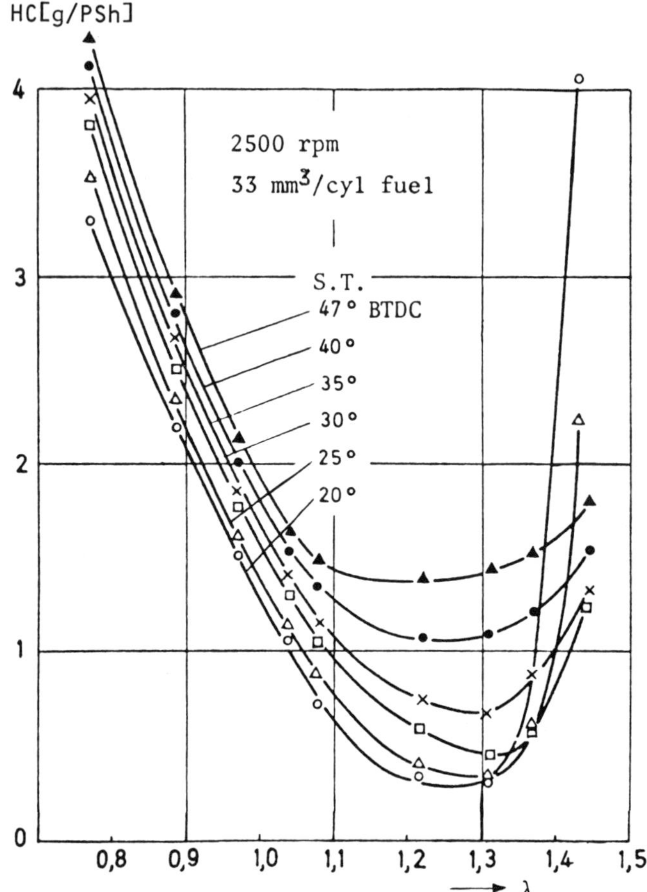

FIGURE 4. Variations in HC Emissions with Spark Timing and Air/Fuel Ratio

We have seen that for high efficiency we should operate with lean ratios at high CR. Why don't we do so?

There are fuel limitations. Knocking limits are compression ratio limits, above which mixtures will ignite due to compression heating and where burning will be explosive rather than smooth. When the engine knocks there is detrimental pounding of the cylinders and bearings.

Ignition limits are mixture ratios which are too rich or too lean for a flame to propagate in a smooth fashion at every firing.

These limits are shown in Figure 8 for an engine operating at wide open throttle. The knock limit for 92 ON fuel, the approxi-

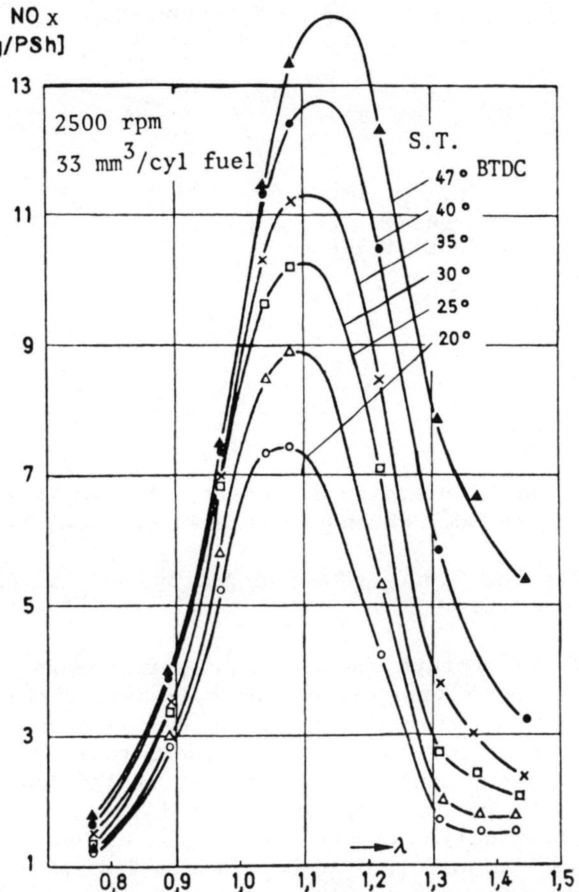

FIGURE 5. Variation of NO_x with Spark Timing and Air/Fuel Ratio

mate value of currently available fuels, is shown. These limits are not exact: they depend on ignition timing, inlet air temperature, moisture level of the incoming air, the quality of air fuel mixing, the composition of the fuel, as well as mixture turbulence. The small arrows show the direction that the knock limit moves as the mixture is made more or less knock sensitive.

In this figure we can see again that the use of the catalytic converter forces operation at the worst possible place, where we can employ only the lowest compression ratio, namely $\lambda = 1$.

In controlling an engine we want to head for the upper right hand corner, in order to achieve high efficiency by high CR, with a

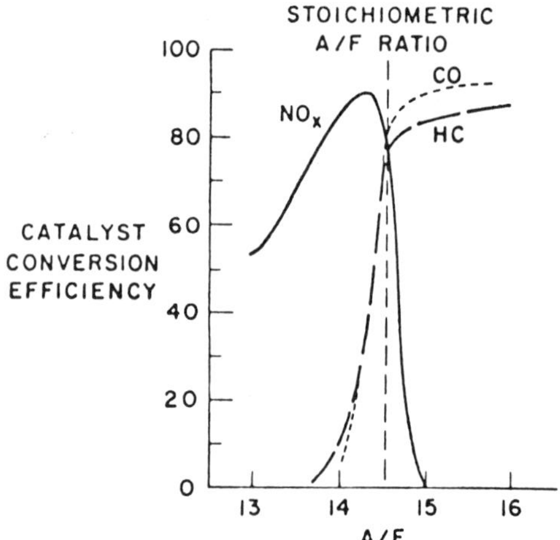

FIGURE 6. Net Conversion Efficiency of a Three-Way Catalyst for NO_x, CO and HC as a Function of Air/Fuel Ratio

given ON fuel. The figure shows we can operate at 12/1 CR with 92 ON fuel at $\lambda = 1.5$.

Why don't we operate there? Lean operation requires good mixture distribution to keep all of the cylinders firing properly all of the time, and to prevent any single cylinder from exceeding the lean limit. The carburetor at the engine intake adds a liquid fuel to the air. The two don't mix well: much of the fuel dribbles down the intake walls on its way to the cylinder while the air gets there almost instantly. Figure 9 shows the effect of the different transport times for air and fuel. In the figure the variation in air/fuel ratio is shown as an engine is accelerated and decelerated, although the steady state air/fuel ratios are approximately constant. With such large dynamic variation in it becomes necessary to choose an average which will not result in excessive misfires on lean mixtures. Since the lean limit is set by the cylinder with the weakest mixture, reduction in the variation of in transient engine operation would permit the lean limit to be extended considerably. To perhaps do this requires improving the mixture distribution to each cylinder.

Using an homogenized gaseous mixture of gasoline and air the lean limits of the engine can indeed be extended. This is shown in Figure 10 where an homogenized air fuel mixture is used. At low engine speeds where liquid fuel mixing is poorer, lean extension of up to 5 A/F ratios can occur. At high engine speeds the lean extension is negligible. It is clear that use of a gaseous fuel is much preferable to that of a liquid fuel in operating an engine.

Look at Figure 8 again. The octane rating of methane is about

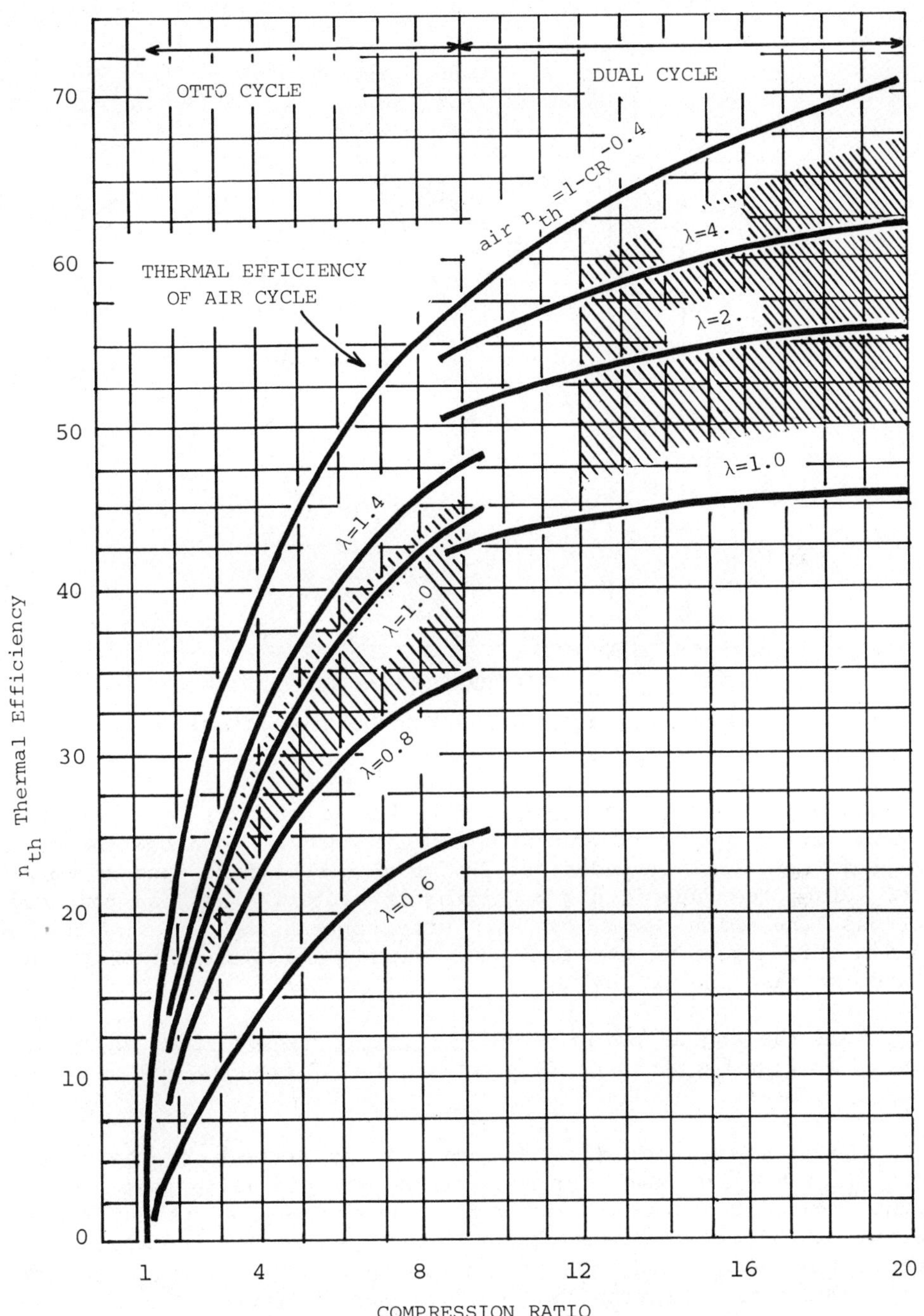

FIGURE 7. Thermal Efficiency vs Compression Ratio and Air/Fuel Ratio

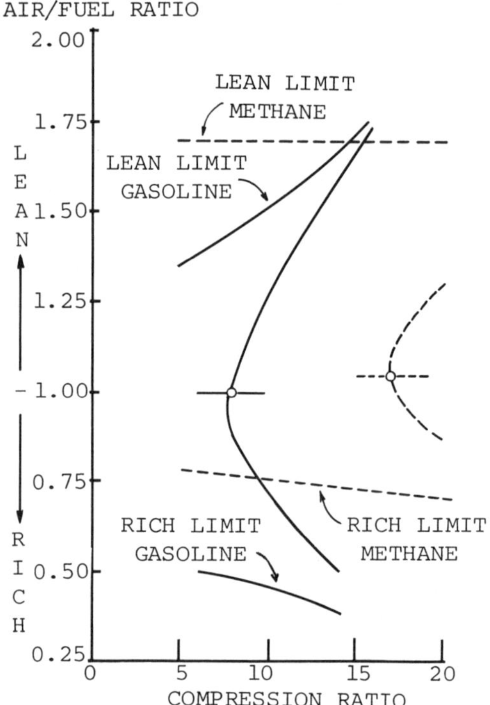

FIGURE 8. Ignition and Combustion Limits

130. The knock limit and ignition limits of methane make possible a much better choice of engine operating point. Note how the lean fueling limit is independent of CR for a gaseous fuel such as methane. Since the fuel is a gas, mixing is better. The lean boundary for gasoline which seemed to vary with compression ratio is not a fundamental limit of the fuel but, rather, is a result of poorer mixing of gasoline at low CR.

From the engine point of view, natural gas provides an opportunity to operate at high CR, with high efficiency, and with low pollution levels.

The fine air fuel distribution one can achieve in the cylinders eliminates the need for much of the paraphernalia that has encrusted current engines. Gaseous fueled engines permit the elimination of chokes, fuel enrichment devices at large throttle angle, accelerator pumps, and catalytic converters. The use of a gaseous fuel reduces oil dilution and prolongs engine life.

Methane is not a poor substitute for gasoline. Methane is a superb fuel.

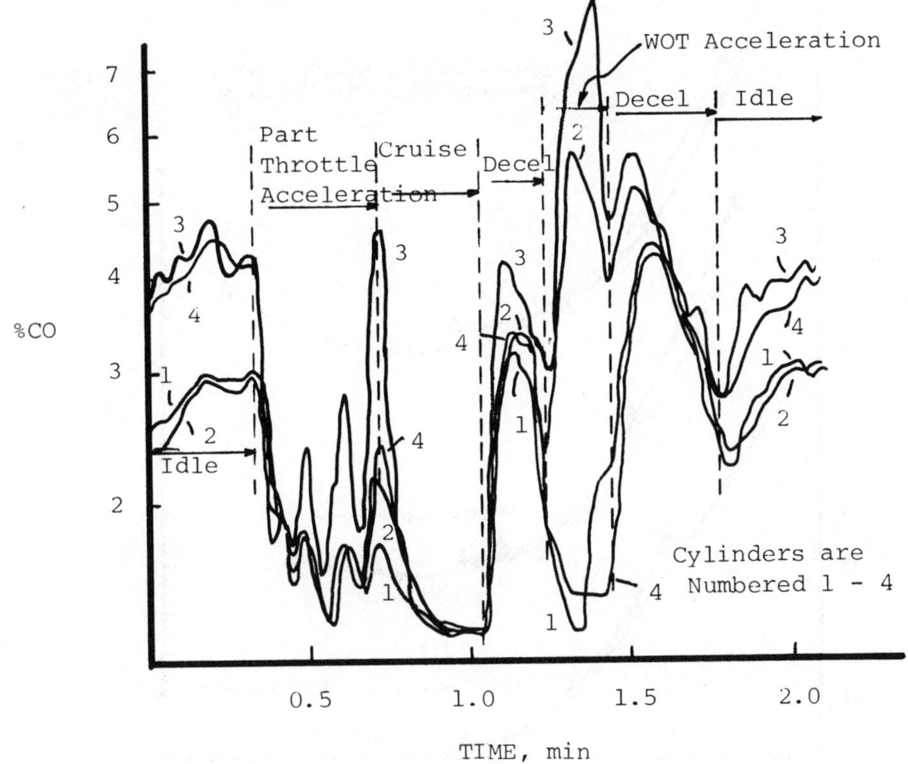

FIGURE 9. The Effect of Different Transport Times for Air and Fuel

About Methane Availability

The world's "Published Proven Reserves" of "Natural Gas" are about the same as those of crude oil - 650 billion barrels of oil equivalent (1979). The consumption rates of natural gas are about 40% of crude. To this we need to add conversion of coal to methane and conversion of biomass to methane. It would appear that methane is a much more available energy base for automobiles than crude oil.

Technical Problems Associated with Natural Gas for Motor Vehicles

Range:

Since gas density is low, methane occupies more space for the same fuel energy as gasoline. Methods of energy density enhancement are needed. Using compression storage, at 200 atmospheres, in a 100 liter tank, in a 1000 kg vehicle, powered by a 1.8 liter engine, we have experienced a range of about 400 km.

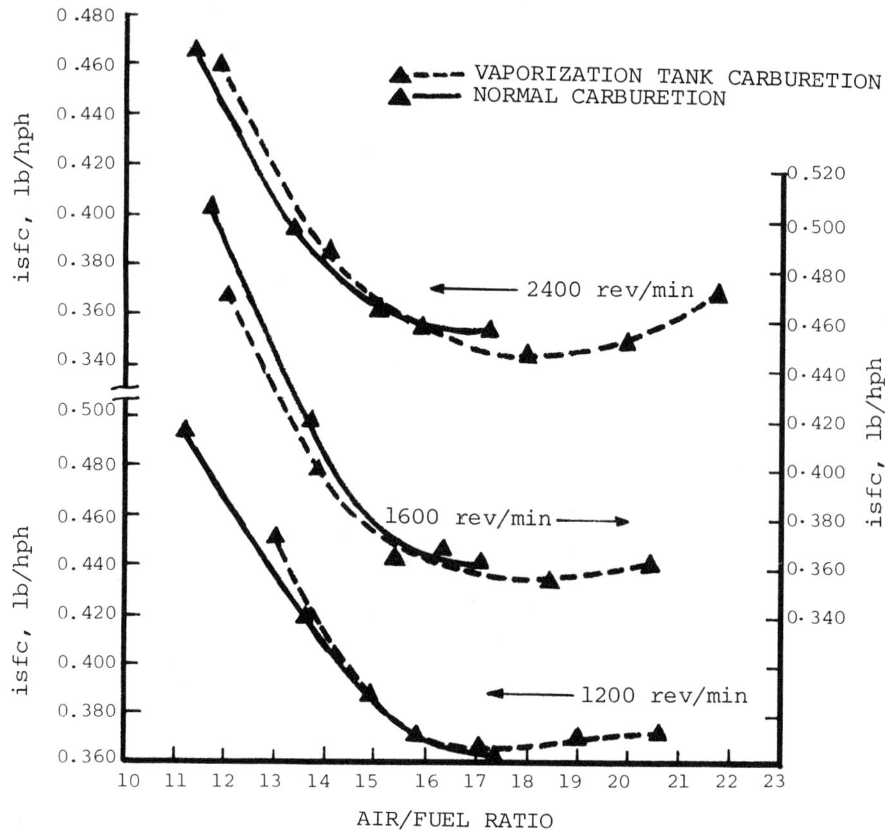

FIGURE 10. Effect of Air/Fuel Ratio on ISFC (Indicated Specific Fuel Consumption) Vaporization Tank and Normal Carburetion
(1 lb/hp h=0.169 kg/MJ; 1 rev/min = 0.1047 rad/s)

Power:

Since the fuel is a gas, methane displaces some intake air thus reducing the effective volume of the engine. This reduces the peak power available. Since methane burns more slowly than gasoline (which gives the high octane number), there is a peak power loss - typically 15%.

Ignition & Timing:

Since the methane molecule is compact and very stable it is hard to ignite. Ignition temperature is about

1170°F compared to gasoline at about 570°F. Better ignition and timing systems are needed.

None of these problems need be solved before methane can be useful as a fuel. Solution of these problems will make it an even better fuel.

The next chapter will tell you briefly about the research program we are conducting to enhance the utility of natural gas and vehicle fuels and to deal with these problems.

THE NATURAL GAS FUELED ENGINE

Gerard J. Born
University of British Columbia

Enoch J. Durbin
Princeton University

Introduction

Currently Canada pays in excess of $7.5 million a day for oil imported from abroad. Use of an indigenous alternate fuel can reduce this national burden. An evaluation of the various options possible indicate that natural gases (NG) offer the most attractive alternative to imported oil as a vehicle fuel source.

Engines have been operated on natural gases since the last century. (General references on the internal combustion engine for gasoline and NG are in the bibliography.) There are limitations in the use of natural gases as a fuel, and these limitations will be diminished by new developments. There are striking advantages in the use of NG as a fuel, and these advantages can be enhanced by new developments. A relatively large number of automobiles, approximately 250,000, are currently fueled by both gasoline and natural gas, among other countries, these vehicles are in Italy, New Zealand, the U.S.A., the United Kingdom, France, and the Netherlands.

A typical automotive conversion for operation on gasoline and natural gases is shown in Figure 1. The gasoline fueling operation remains unchanged. The natural gases fueling system is added. This consists of storage tanks, pressure regulators, a gas/air mixer, a fuel selector, fuel switching mechanisms, and a NG refueling connector.

Engine Power

The power and power losses of an engine can be measured from a

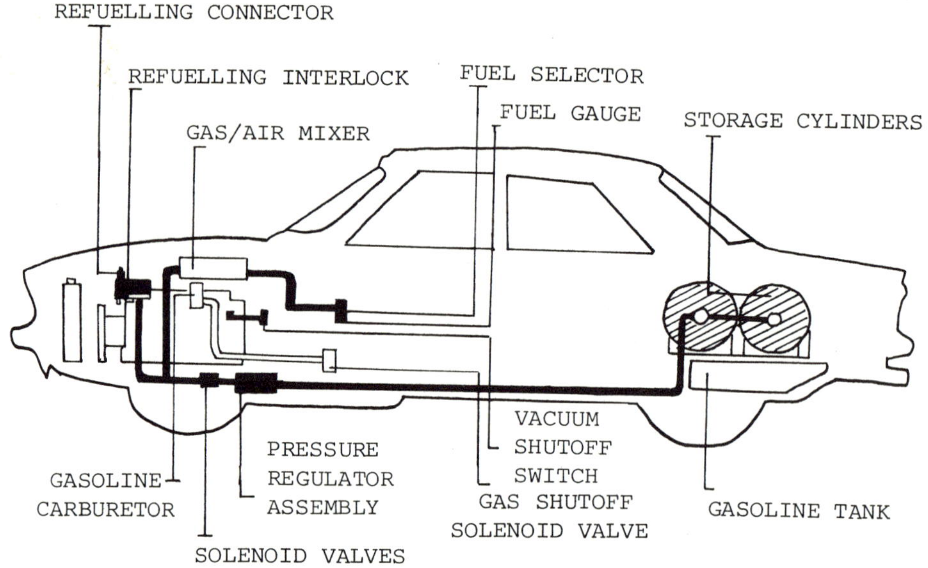

FIGURE 1. Alternate Fuel Conversion (CNG-Gasoline)

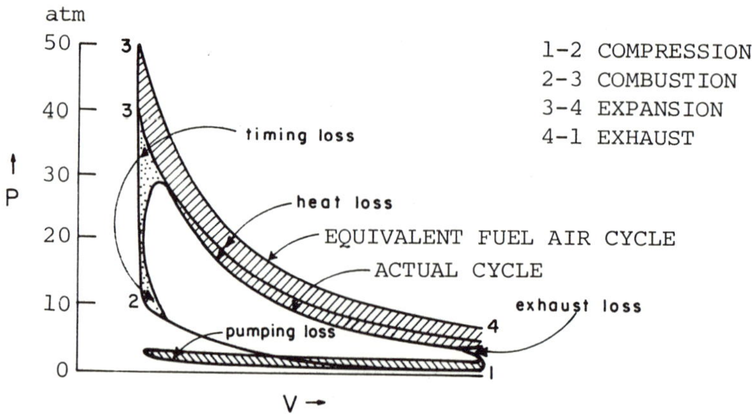

FIGURE 2. Power Loss in a Spark-ignited Engine

pressure volume (PV) diagram, as shown in Figure 2.

In an idealized PV diagram the cycle starts with the cylinder filled with an air/gas mixture at a pressure and volume as indicated by point 1. The piston compresses the mixture adiabatically until it reaches top dead center, point 2. The gas mixture is burned instantaneously and causes a pressure rise, point 3. The expanding gas expands adiabatically to the original volume, point 4, followed by an instantaneous exhaust to the starting point, 1.

In an actual engine cycle there is:

a) pumping loss (intake and exhaust gas pumping energy);
b) timing loss (mixture is not burned instantaneously);
c) heat loss (heat is not converted into work);
d) exhaust loss (hot, still pressurized gas is exhausted at the opening of the exhaust valve), and
e) friction loss (piston rings, bearings).

The methane flame speed is slower than the gasoline flame speed, hence the timing loss for NG operation is larger than for gasoline.

The approximate engine losses are shown in Figure 3 (reference 6). This figure shows an overall efficiency of about 16%. This picture represents the dynamic operation of the engine. Under steady state conditions the engine output efficiency rises to about 28%.

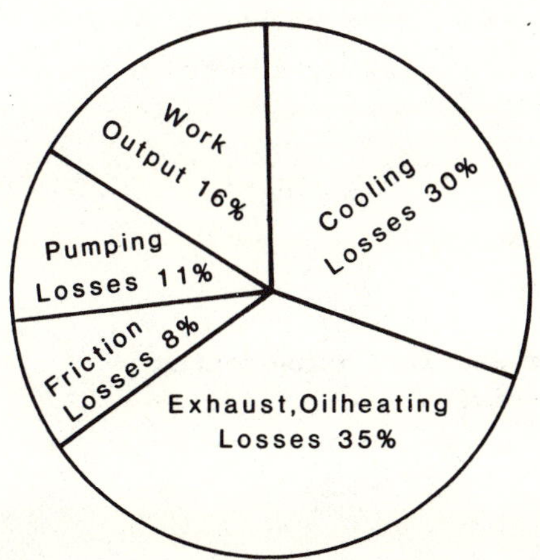

FIGURE 3. Distribution of Fuel Energy in Road Operation (North American Automobile 1981)

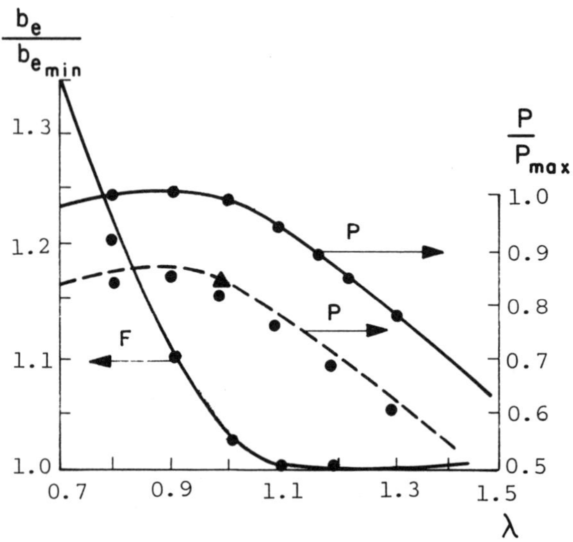

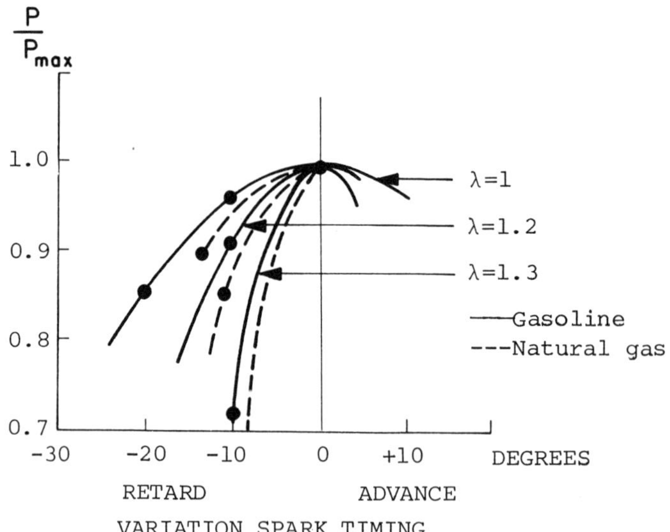

FIGURE 4. Normalized power and fuel consumption and timing sensitivity.

In an engine the gasoline or NG is mixed in the carburetor or gas mixer. The chemically correct air/fuel ratio, or stoichiometric mixture, is 14.7 for gasoline, and 17.2 for NG operation.

Lambda is defined as the actual air/fuel ratio divided by the stoichiometric ratio. Lambda equals 1 represents a stoichiometric ratio, a lambda larger than 1 corresponds to a lean ratio, and a lambda smaller than 1 depicts a rich ratio.

The normalized power P/P max is the ratio of the actual power divided by the maximum power.

Figure 4a shows the normalized power as a function of lambda.

In these experiments the air intake is held constant and the amount of fuel is varied. Therefore the fall-off in power at lean mixture is due to the reduction in fuel input.

The maximum combustion temperature does not occur at the stoichiometric mixture but, due to dissociation and specific heat variation, it occurs at a slightly rich mixture. The maximum power is thus obtained at a slightly rich mixture.

At maximum engine power, the throttle is fully opened, hence maximum power available is reduced at lean air/fuel ratios due to the smaller fuel flow. The loss in maximum power for NG is about 10 to 15% at 2500 rpm. This power loss is caused by three factors:

1. About 8% of air is replaced by NG in the air intake. This is a higher percentage than with a liquid or more dense fuel.

2. The flame speed of NG is slower than that of gasoline, hence the timing loss is larger.

3. The evaporation of a liquid fuel in the engine air intake cools the incoming air and raises its density, increasing the engine's air mass breathing capacity.

The effect on power due to a variation of the spark timing is shown in Figure 4b. Note especially the relative insensitivity of power to spark timing with rich mixtures, and the extreme sensitivity at lean mixtures. Traditional operation of spark ignited engines with rich mixtures has permitted a rather imprecise control of spark timing to be quite satisfactory.

In lean operation good air/fuel ratio and timing control is necessary.

Engine Fuel Economy

A measure of the fuel economy is fuel consumption, per horsepower hour, b_e. This is normalized by the minimum fuel consumption per horsepower hour b_e min.

Figure 4a shows the normalized minimum fuel consumption as a function of air/fuel ratio, lambda, for a natural gas powered engine and for a gasoline powered vehicle.

The thermal efficiency is defined as the ratio of work per cycle to the heat released by fuel combustion. When the air/fuel mixture becomes leaner (lambda greater than 1), the thermal efficiency increases. For the same amount of fuel at a leaner mixture more useful work can be delivered. Therefore, the minimum specific fuel consumption occurs at lean mixtures.

Note the large variation in specific fuel consumption when the fuel/air ratio is varied.

In the same engine the normalized specific fuel consumption for NG operation is practically the same as that for gasoline. The effect of a variation of spark timing on fuel consumption is similar to that shown in Figure 4b. Fuel consumption is relatively insensitive to spark timing at a rich mixture, but is extremely sensitive to spark timing at a lean mixture.

The timing for NG fuel is different than that for gasoline. At 2500 rpm the following best timing angles were measured for NG and gasoline.

λ	1	1.1	1.2	1.3
Gasoline	20°		30°	45°
Natural Gas	32°	35°	40°	50°

Optimizing Power and Fuel Economy

Under driving conditions, a specific amount of power is required for a specific vehicle, speed, or acceleration. This power can be provided at a variety of air/fuel ratios, with varying fuel economy.

The determination of best operating point is most clearly made by an experiment in which a constant fuel flow rate is provided and the air quantity is changed by changing the throttle (5). Such data permit one to see the changes in fuel economy and power that the engine can provide at different operating conditions.

Engine power and fuel consumption as a function of λ are shown in Figure 5. Both variables are normalized for stoichiometric air/fuel ratio (lambda = 1.0), and the amount of fuel is kept constant, while the air intake is varied.

For a given amount of fuel the available engine power increases at lean air/fuel mixtures due to increased thermal efficiency.

Note the sensitivity of fuel consumption to air/fuel ratio. Measurements in carburetors and gas mixers both indicate a large air/fuel ratio variation during actual driving conditions. Typical air/fuel ratio variations are in the order of lambda 1.2 to 0.8. A considerable saving in fuel can be obtained if one can provide a constant and lean air/fuel ratio.

Pollutants

The dependence of engine pollutants on λ and timing is shown in Figure 6. Note that pollutants for the methane-fueled engine are lower for the same λ than for gasoline.

Figure 6a shows the variation of CO with air/fuel ratio and spark timing. CO depends almost exclusively on the availability of sufficient O_2 to complete the conversion of CO to CO_2. At rich mixtures there is insufficient O_2.

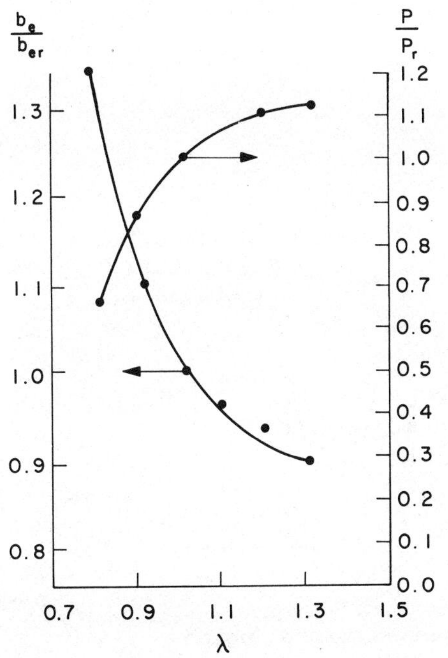

FIGURE 5. Power and Fuel Economy Normalized at Lambda = 1 (Fuel Flow Constant)

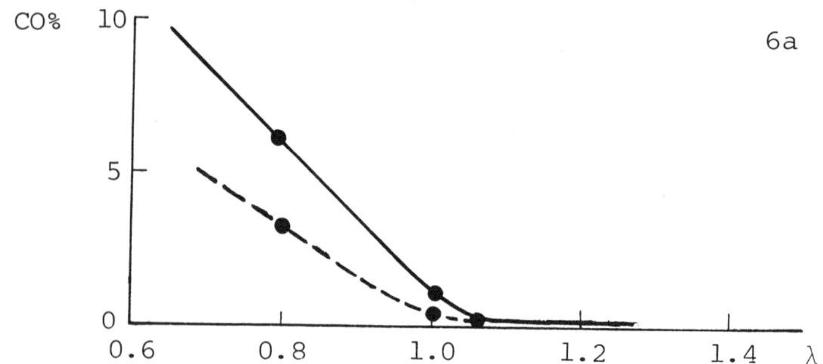

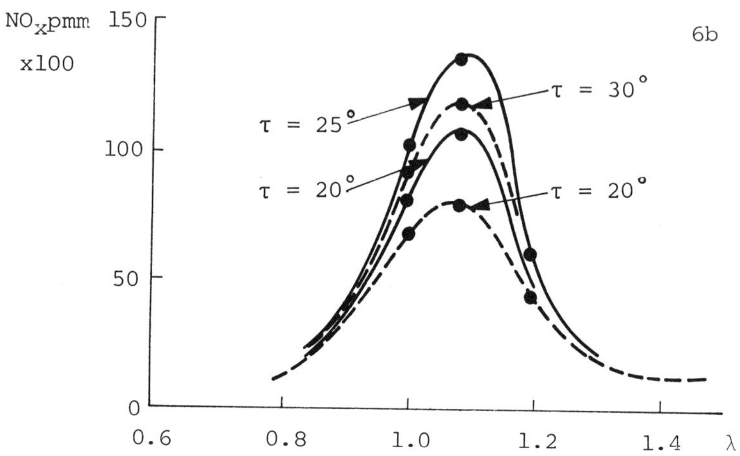

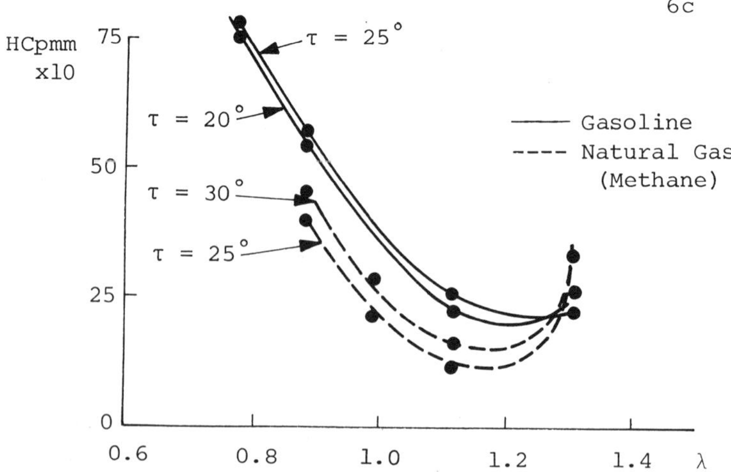

FIGURE 6. Pollutants in Spark-ignited Engine

Figure 6b shows the variation in the rate of production of NO_x (various oxides of nitrogen) with air/fuel ratio and spark timing. NO_x production is a non-equilibrium process. The quantity of NO_x produced from a given quantity of fuel depends on the temperature/time history of the mixture and the availability of O_2. No_x is strongly dependent on maximum temperature reached. Lower NO_x levels at rich mixtures occur because there is a deficiency of O_x and lower combustion temperatures. The temperatures are lowered by the absorption of some of the heat of combustion by the excess fuel.

Low NO_x at lean mixtures occurs because the combustion temperatures are lowered by the absorption of some of the heat of combustion by the excess air. Still lower levels of NO_x would be observed at lean mixtures if the excess air is replaced by exhaust gases (mainly CO_2 and H_2O), thus reducing the O_2 available for oxidation of N_2 and lowering the combustion temperature by increasing specific heat.

NO_x production is relatively insensitive to spark timing with rich mixtures. Increased NO_x at stoichiometric and lean mixtures as the spark timing is advanced occurs because of the higher peak temperatures reached when the mixture is ignited early and then compressed by the rising piston. This further increases the temperature of the burnt gas.

It is important to note that, because of the non-equilibrium character of the NO_x formation process, the portion of the fuel which is first burned contributes more to the formation of NO_x than the portion which is later burned. To obtain lower NO_x levels, without sacrificing fuel economy, requires that the combustion propagate quickly after ignition, reducing the required spark advance.

A smaller spark advance reduces the duration of the high temperature and reduces the adiabatic compression which the first burned portion of the mixture experiences. This effect can be achieved by improved ignition sources, better mixing, and more turbulence in the cylinder mixture.

Figure 6c shows the variation of HC (hydrocarbon) with air/fuel ratio and spark timing. HC levels are the resultant of three effects acting together. Lower combustion temperatures with rich or lean mixtures increase the cylinder quench distance, the region of the cylinder near the cooled engine surface where the mixture temperature drops below its ignition temperature. This causes part of the rise in HC on both sides of stoichiometric where the combustion temperature is a maximum. A large part of the rise at the rich end is due to fuel being provided in excess of that which can be oxidized by the available O_2.

The sensitivity of HC production to spark timing is most pronounced with lean mixtures. HC is less sensitive to spark advance at rich mixtures. Retarded spark timing causes the mixture to burn very late in the combustion cycle raising the exhaust gas temperature. This permits oxidation of the residual hydrocarbons in the exhaust system.

The rise in hydrocarbons at very lean mixtures occurs because the mixture temperature during the expansion stroke is permitted to drop below the ignition level before the burn is complete. Faster burning by better ignition and heightened turbulence can delay this HC lean rise to still leaner mixtures.

For low engine pollutants a good air/fuel ratio control, with appropriate timing, is necessary.

Vehicle Range

One of the deficiencies of NG powered vehicles is limited range. The range of a vehicle is determined by:

1. The fuel and its properties, (gasoline, NG, and methane-propane mixtures).
2. The amount of fuel stored.
3. The fuel consumption rate, which is a function of the power required to operate the vehicle, fuel properties, and engine efficiency.

Engine efficiency depends on compression ratio, turbo or supercharging, air/fuel ratio, and spark timing.

The range can be increased by improvement in each of the above factors, and each of these is being studied in the Alternate Fuels Laboratory of the University of British Columbia.

One research effort concentrates on extending the vehicle range by improvement of the fuel energy density in the storage system and by alternate storage concepts, such as storage by solution, absorption on powdered solids, and occlusion on other materials.

One of the storage concepts proposes to dissolve methane in a solvent to increase energy density (7).

The feasibility of operating an engine on propane-methane mixtures was demonstrated in a high compression ratio engine (12:1), which was displayed at the international conference. Both fuels were stored together in one tank. This concept has led to improvements in range. A change in compression ratio from 8:1 to 12:1 can give 20 to 25% improvement in fuel economy. A methane-propane mixture with 10% propane should give a range increase of about 20 to 25%. The total range increase is the sum of both improvements, since they are both independent of each other.

Research Program at the UBC Alternate Fuels Laboratory

The goal of the Alternate Fuels Laboratory is the research and development of the alternate fueled vehicle to improve the efficiency, convenience and safety of natural gas as a transport fuel.

Research efforts are on the control and performance of motor vehicle engines fueled by natural gas, and improvement of vehicle range.

One effort focuses on an investigation of an accurate air/fuel ratio control system. Another effort concentrates on ignition sources and ignition timing control, and a third effort is concentrated on range extension by energy density enhancement. In this connection it is quite possible that future gaseous vehicles may have to operate on varying fuel composition, hence it may be necessary to employ feedback air fuel controllers.

Conclusion

Although there are many significant research opportunities in enhancing NG fueled vehicles, implementation of a program to convert vehicles to NG operation does not require the results of this research to appear. The performance of vehicles operating with existing equipment is satisfactory. Research will make such performances even better.

REFERENCES

1. West, J.P., Brown, L.G., "Compressed Natural Gas", New Zealand Energy Research and Development Committee Publication, p 14, April 1979.

2. Taylor, C.F., The Internal Combustion Engine in Theory and Practice. Vol. I, II, MIT Press, 1968.

3. Obert, E.F., Internal Combustion Engines and Air Pollution; Harper & Row, Publishers, 1968.

4. Benson, R.S., and Whitehouse, N.D., Internal Combustion Engines, Permagon Press, 1979.

5. Zeilinger, K., "Beitrag zur Untersuchung der Schadstoffemissionen eines Ottomotors unter besonderer Berucksichtigung des instationaren Motorbetriebes", (Investigation of Harmful Exhaust Gas Emissions of Spark Ignited Engine Specially under consideration of Unsteady Engine Operation), Ph.D. Thesis, Munich Tech. Univ., 31 July 1974.

6. Private Communication, General Motors Laboratory.

7. Private Communication, Dr. R.O. McElroy, B.C. Research.

8. Alternate Fuels Laboratory, University of British Columbia, Report No. AFL 81-01 (to be published).

METHANE AND DIESEL ENGINES

G.A. Karim

Department of Mechanical Engineering

University of Calgary

In other chapters you will read that diesel engines can be operated on gaseous fuels. Accordingly, I shall try to bias my presentation towards some features associated with such operation and emphasize the comparative aspects of operating diesel engines on methane relative to propane, hydrogen or ethylene because this is an issue that needs to be addressed.

The so-called dual-fuel engine, that is the gas diesel engine, is not new. In fact, the first patent that was given to operate engines on the dual-fuel principle went to Rudolph Diesel (1) who was not very successful in its operation as the engine knocked very badly. Later the Germans tried, without success, to operate diesel engines on hydrogen. The first commercial application of a dual-fuel engine was in the thirties and during the second world war much effort was made to use poor quality gasoline in the form of gasified vapour to run in diesel engines as well as coal gas and methane. After the war, interest in dual-fuel engine applications went up and down depending on the price of gaseous fuels and the extent of competition from other fuels. However, most of these efforts can be described as having been made on an *ad hoc* basis and relatively little basic research was done to solve some of the problems normally encountered. Even nowadays, unfortunately, there still appears to be a gap between those who do engine conversions and the mass of technical information available in the literature.

There are many diesels that are operating on methane and other gaseous fuels. These are mostly for stationary applications because they lend themselves extremely well for such applications, particularly where a cheap and fluctuating source of fuel gas is available. Operation can, for example, be made primarily on natural gas until the supply of gas is either depleted or interrupted when a changeover to pure diesel operation can be made without loss of power or loss of speed. In fact, many sewage works utilize diesel engines that operate on sewage gas, a mixture of methane and carbon dioxide for the production of power.

Obviously, to run diesel engines on natural gas you will get most of the benefits of the diesel and they are many. For example, diesel engines are highly efficient, amenable to turbocharging and have the right torque versus speed characteristics. They are very good on maintenance, have a long operational life and they have good emission characteristics. Also, in many parts of the world diesel fuel is relatively cheaper than other liquid fuels.

What I will do before discussing some features of engine performance on methane is to review some aspects of the combustion

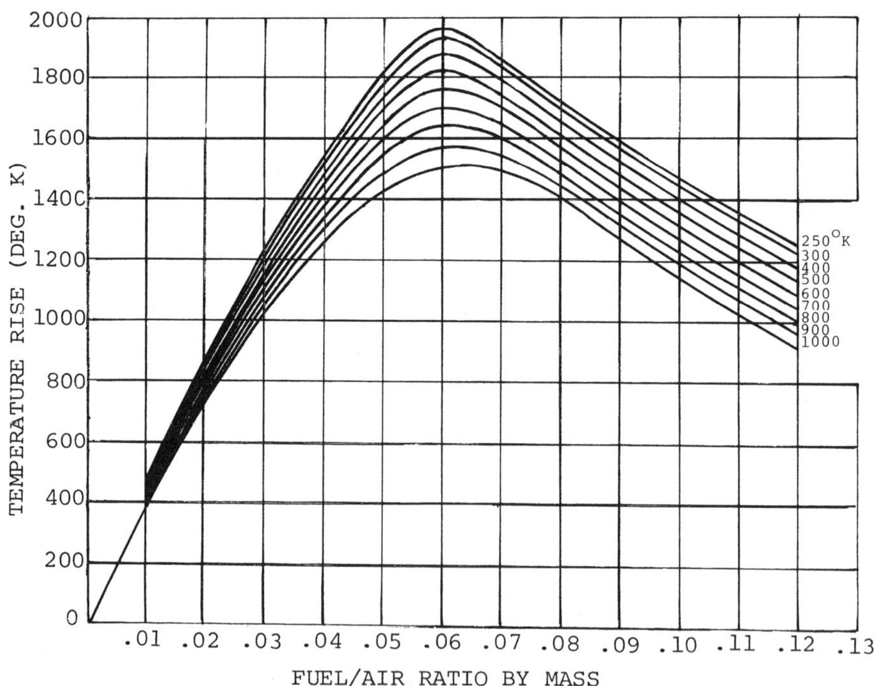

FIGURE 1. The temperature rise following the adiabatic combustion of methane-air for a range of initial mixture temperature and fuel to air ratio (2).

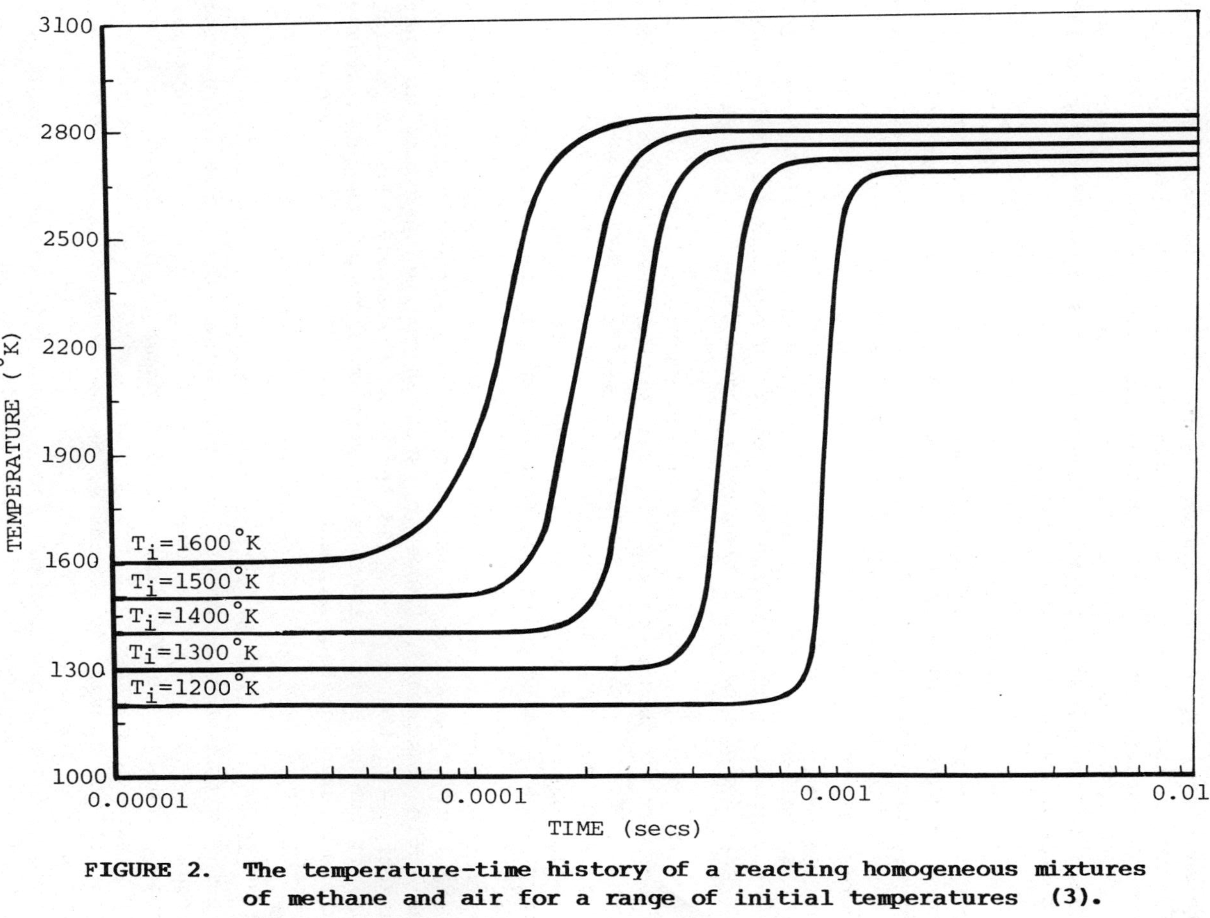

FIGURE 2. The temperature-time history of a reacting homogeneous mixtures of methane and air for a range of initial temperatures (3).

of methane that have been referred to by other authors. Figure 1 shows the temperature rise following the combustion of methane in air versus the fuel-air ratio (2). It can be seen that for the lean methane-air mixture range over which engine operation normally takes place, there is a linear dependence of the temperature rise following combustion on the fuel-air ratio supplied. In other words, unlike quantity control in spark ignition engines, controlling mixture quality can control the amount of energy supplied to the diesel engine. However, these data do not take into consideration the effect of time.

Figure 2 relates to a methane-air mixture in stoichiometric proportion initially being allowed to react at a certain temperature so that eventually it will autoignite after a certain period of reaction time (3). This time required for autoignition can be seen to be logarithmically dependent on temperature. In other words, if temperature is lowered by a small amount, the time required for autoignition increases very significantly. Therefore, the so-called ignition temperature of methane is an indefinite term that depends on many parameters including how long we are prepared to wait for the fuel-air mixture to autoignite.

Therefore, in the diesel engine where we have got a limited time, of the order of one or two milliseconds around top dead center, for the fuel-air mixture to ignite, a much higher temperature than normally encountered in the compression process of the diesel engine is needed to autoignite the methane and air. In fact, we have autoignited methane and air in an engine without any diesel fuel injection by increasing the engine compression ratio to extremely high values, in excess of 20:1, or preheating the air to very high temperature so that the fuel-air mixture reacts actively as it is compressed. However, such an engine operation is unattractive and the engine would knock very severely (4). Accordingly, it can be said that in a conventional diesel engine, methane-air mixtures cannot be made to autoignite on their own in an acceptable manner.

If a fuel like diesel or diesel vapour is introduced into a region within the methane-air mixture then the temperature range required for local ignition there will be depressed significantly, leading to autoignition. The flame that may propagate from such an ignition zone, with hope, can consume the rest of the mixture irrespective of how lean it is. However, this may not be the case. The ignition limits (4, 13) of methane-air mixtures when spark ignited for a very wide range of compression ratios and engine intake temperatures shown in Figure 3 are relatively narrow. It can be seen that the lean ignition limit does not change very significantly as the compression ratio is changed in this case from 8:1 to 16:1. However, the ignition limit appears to be very sensitive to changes in temperature. This is

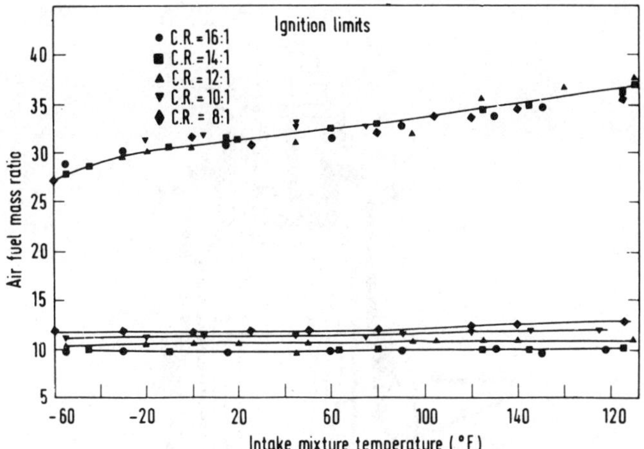

FIGURE 3. **The ignition limits of methane-air mixture in a spark ignition engine for a range of intake mixture temperatures and compression ratios (13).**

why we can get adequate combustion of methane following the injection of some diesel fuel. However, should a too lean mixture be employed with a relatively small diesel pilot injection, flames initiated from the ignition centers of the diesel pilot cannot propagate throughout the whole length of the cylinder to consume the entire gaseous fuel-air mixture. This is one of the problems confronting diesels when operating on dual fuel and very light loads (5).

Figure 4 shows a schematic diagram of a marine diesel engine that operates on methane derived from the boil-off of LNG (6). It is a two-stroke engine of a very large bore that is being used in some tankers transporting liquid natural gas from the Persian Gulf or Indonesia to Japan. The engines run on some of the boil-off of LNG. On the return journey the engines are operated as conventional diesels. This is a significant achievement involving two-stroke, turbocharged and very large bore engines.

The laboratory engines that we have converted to run on diesel and gaseous fuels are quite typical and the conversion is very straightforward. The gaseous fuel is simply carbureted or introduced into the air (7). The effect of low temperature operation can also be simulated by jacketing the engine intake manifold with liquid nitrogen (8). Thus, operation on LNG boil-off can also be simulated. There are some industrial applications, particularily with turbocharged two-stroke engines, where timed injection of the gaseous fuel directly into the cylinder is needed requiring

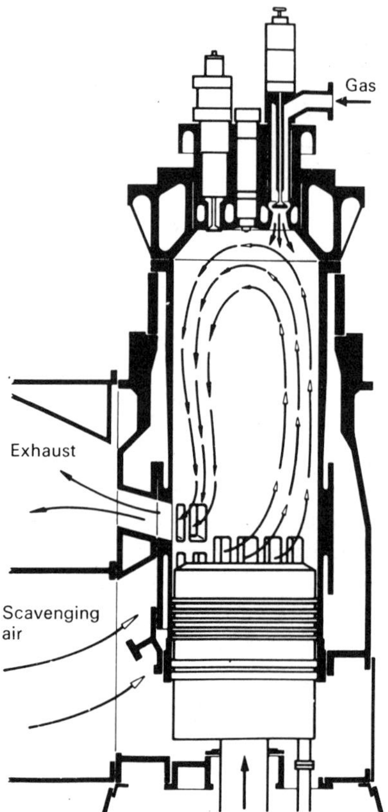

FIGURE 4. Gas injection for Sulzer crosshead-type dual fuel marine engine (6).

some major and relatively expensive conversion.

Great care is needed when comparing the performance of engines on different fuels such as gasoline versus methane operation or propane versus methane. The comparison must be made for the same engine setup and under exactly the same operating conditions. Unfortunately, some statements can be seen in the literature referring to the relative performance of engines that are not made under the same conditions nor for the same engine. Figure 5 shows a typical comparison between spark ignition and dual fuel operations on natural gas for the same engine (9). It can be seen that the specific fuel consumption (i.e. how much energy is required a unit amount of work) is less for the compression ignition diesel than that for the spark ignition engine. Thus, usually better performance can be obtained with diesel on CNG or other gaseous fuels rather than the corresponding spark ignition mode. The main reason for this superiority in performance, as shown in Figure 5, is that

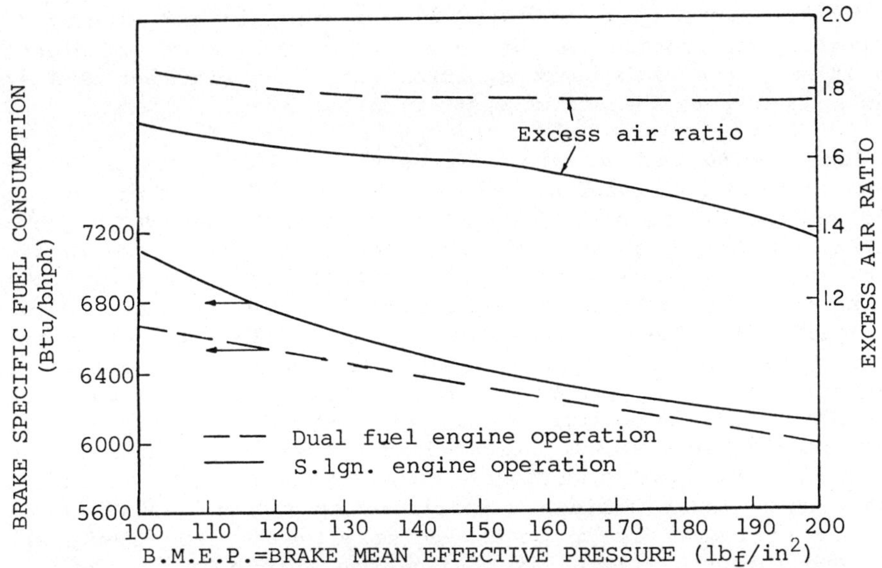

FIGURE 5. **The brake specific fuel consumption and feed excess air ratio variations will brake mean effective pressure of an engine when operated, in turn, as a gas engine and a dual fuel diesel engine on methane (9).**

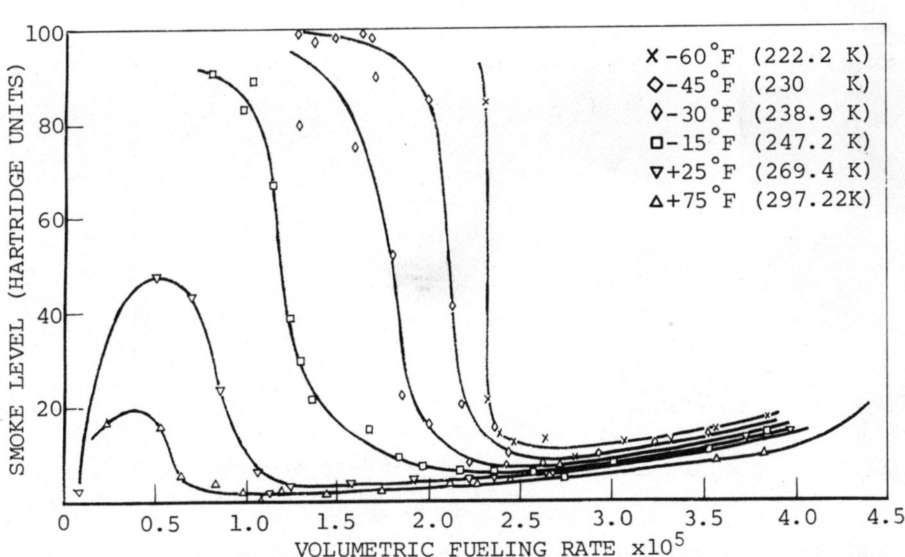

FIGURE 6. **The exhaust smoke density, in Hartridge Units, variation with diesel fueling rate for a range of intake temperature (10).**

the diesel allows operation with much excess air and, therefore, the combustion process can be carried out much more efficiently than in the case with spark ignition where the mixtures used are somewhat nearer to the stoichiometric ratio.

Now, coming back to the advantages and disadvantages of the conventional diesel engine; one of the main limitations of the diesel engine is the fact that the power output usually is smoke-limited. Figure 6 shows typically the exhaust gas smoke density versus engine fueling rate all the way from zero fuel input, when the engine is being motored, right through to idling and almost to full load (10). The various curves are for different ambient temperature operation. It can be readily seen that as the intake air is lowered the exhaust smoke emission becomes a very serious problem. There will be a limit to how much power can be produced for an acceptable smoke level. Light load operation can equally produce unacceptable bad conditions. Such tendencies will worsen even further with lower ambient temperature operation. However, as shown in Figure 7, if the same engine with the same compression ratio and injection timing is operated on a gaseous fuel such as methane, using a small amount of liquid diesel pilot, hardly any exhaust smoke will be produced. Much additional power to that of the pure diesel operation can be produced if needed. It can be

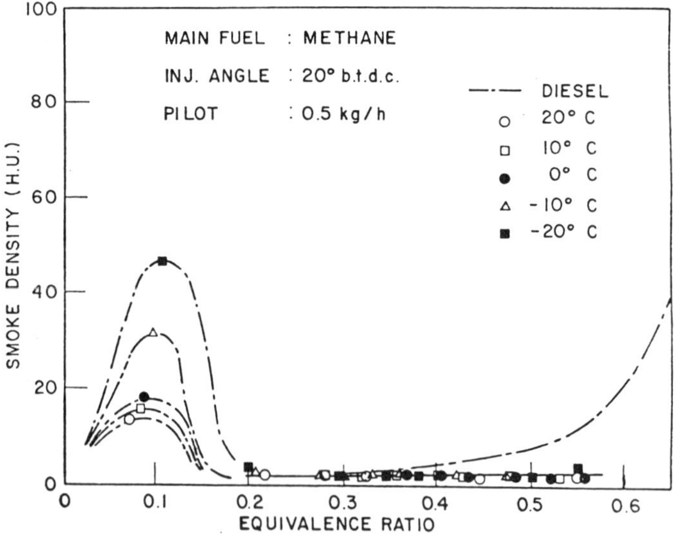

FIGURE 7. The exhaust smoke density in Hartridge Units variation with total equivalence ratio for a dual fuel engine operating on methane for a range of intake temperatures; the diesel case is also shown (8).

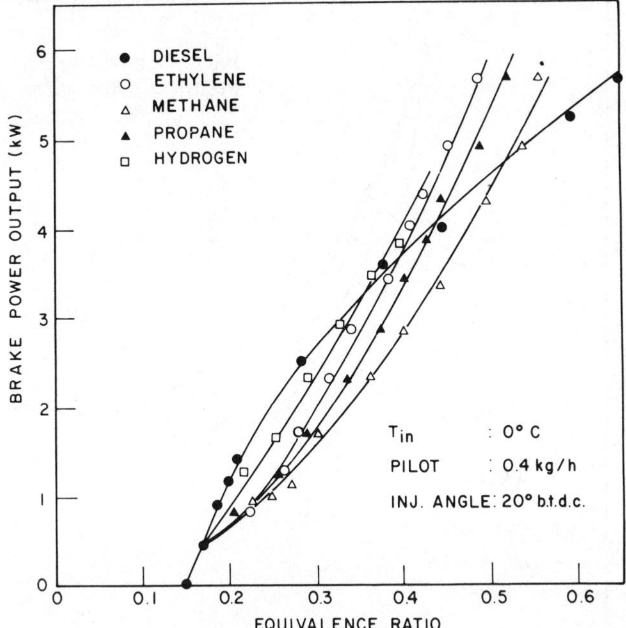

FIGURE 8. Brake power output variation with equivalence ratio of a dual fuel engine when fueled, in turn, with a range of gaseous fuels at constant liquid pilot quantity (8)

seen that even at an intake temperature of -20°C with methane induction, there appears to be no problem associated with the extent of exhaust smoke. Therefore, a very dramatic benefit associated with the gas-diesel operation is the fact that it can be extended to a very high load without encountering significant exhaust smoke. Moreover, such a problem is also absent at very light load.

Figure 8 shows typically the brake power output against equivalence ratios, the total fuel-air ratio relative to the corresponding chemically correct ratio, for a number of cases involving a range of gaseous fuel addition to the engine intake that includes methane, propane, hydrogen and ethylene at constant diesel pilot injection and speed (8). The solid line represents diesel operation. As one would expect, the more fuel is added the more power is produced. It can be seen that at very light load, diesel operation is superior producing slightly more power output for the same amount of fuel. However, when relatively higher fueling rates are employed, diesel operation begins to deteriorate, because of the longer combustion, relative to gaseous fuel applications where very significant increase in power can be obtained well beyond that of

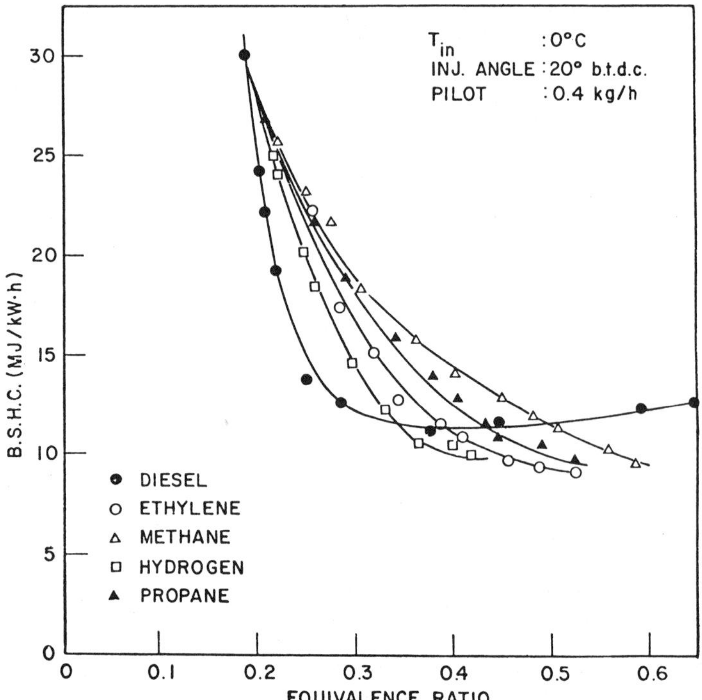

FIGURE 9. The brake specific heat consumption (BSHC) variation with equivalence ratio of a dual fuel engine when fueled, in turn, with a range of gaseous fuels at constant liquid pilot quantity (8).

the conventional diesel. Methane tends to produce slightly less power than propane while hydrogen and ethylene because of their faster burning rates, would be superior to propane.

The corresponding brake-specific heat consumption, being a measure of how much energy is required to produce one unit of useful work is shown in Figure 9. It can be seen that dual fuel operation at light load, irrespective of what gaseous fuel is being used, is slightly less efficient than the corresponding diesel operation. Beyond half-load however, the efficiency of dual fuel operation improves sufficiently later to surpass that of the diesel. It can be clearly seen that methane is much superior to the diesel at full load. Moreover, as far as emission is concerned, oxides of nitrogen from diesel engines tend to increase with the amount of fuel being used almost linearly. However, with the addition of gaseous fuels, the production of oxides of nitrogen can be delayed. Increasing the relative amount of diesel to the methane, more oxides of nitrogen will be produced indicating the need for

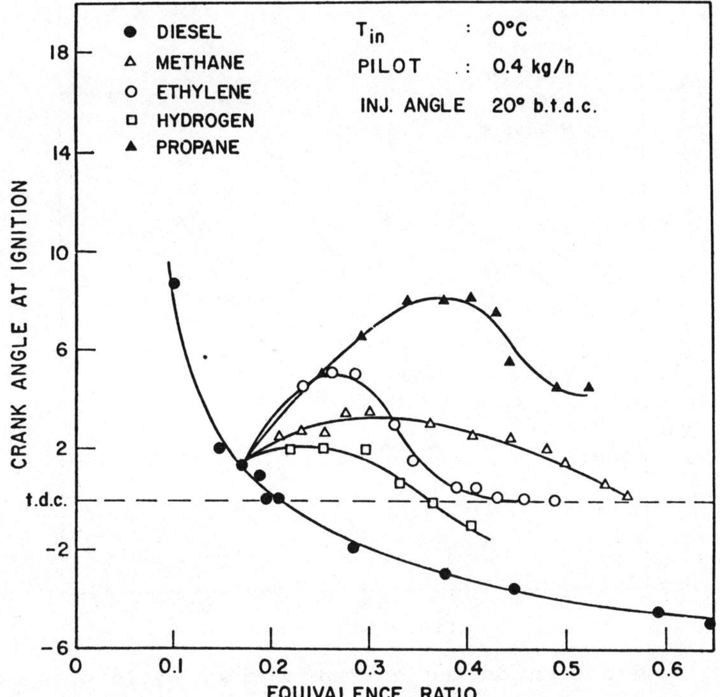

FIGURE 10. Variation of the ignition point with equivalence ratio for a dual fuel engine fueled, in turn, with a range of gaseous fuels at constant liquid pilot quantity and injection timing. Diesel operation is also shown (8).

using a very small amount of pilot fuel (5).

A problem associated with the dual-fuel engine is the relatively poor light load and idling performance. This is, as shown in Figure 10, when a gaseous fuel is introduced. The point at which ignition takes place is delayed thus undermining performance. It is only when the concentration of the fuel is increased significantly that improvement in the ignition delay period can be obtained. This trend is observed with all the fuels used. Methane is superior to propane in this regard. This is believed a consequence of the competition for active radicals between the diesel fuel and the propane affecting the combustion process adversely; thus having the same effect as making the diesel fuel of a much lower cetane number than what is is. It is only when a considerable amount of propane is added that it will generate a significant amount of energy helping the ignition process of the diesel fuel and reducing the ignition delay. But for methane, even with about three or four percent of the gas in the air, the

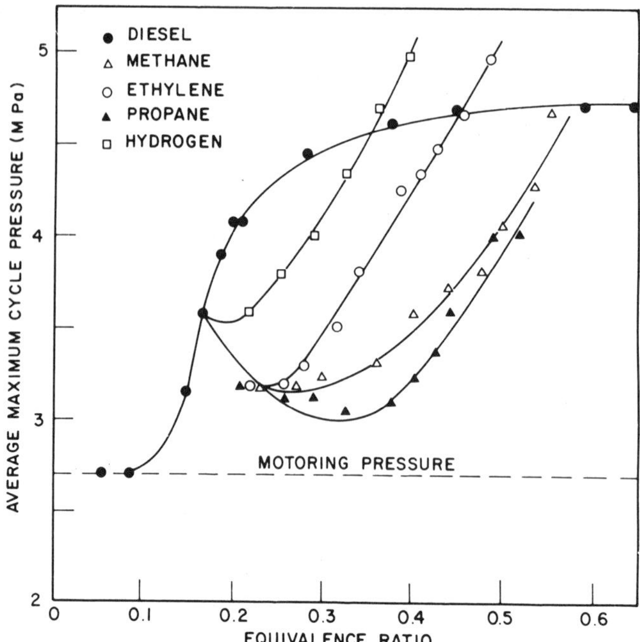

FIGURE 11. **Variation of the average maximum cycle pressure with equivalence ratio of a dual fuel engine fueled, in turn, with a range of gaseous fuels (8).**

ignition delay is comparable to the corresponding pure diesel operation.

Figure 11 shows typically the average maximum cycle pressure when operating on various gaseous fuels. It can be seen that the peak cylinder pressure is reduced considerably with gaseous fuel addition, contributing towards quieter engine running than with diesel. Similarly the corresponding rate of pressure rise is much lower. This is a very definite plus for dual fuel operation, particularly at low temperatures.

At very light load, the low concentrations of methane added to the intake will not necessarily burn completely despite the presence of much excess air. The reason for this is that with the very small amount of methane added, even at the very high temperatures encountered near the peak of compression, and even after the ignition of the diesel pilot releasing energy and raising temperature, the concentration of the methane is so small that the flames starting from these pilot ignition centers will be quenched. Thus, much of the methane can remain unreacted surviving to the exhaust stage. Only by going to slightly beyond the idling range that much

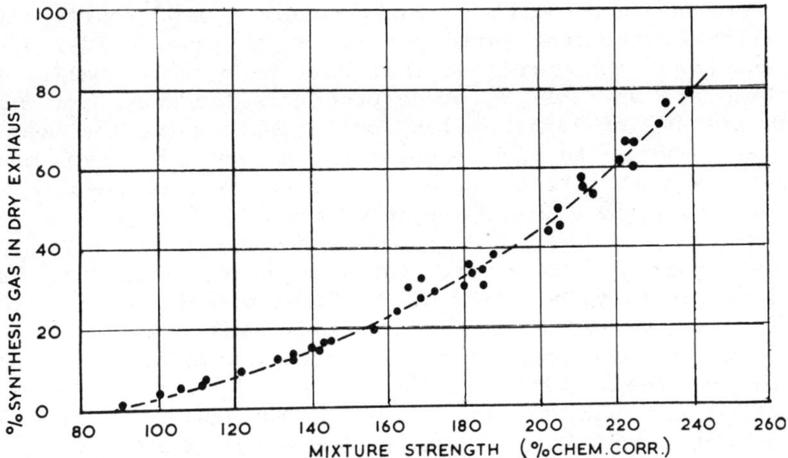

FIGURE 12. The extent of production of synthesis gas ($CO + H_2$) in the exhaust of a dual fuel engine fueled with rich mixtures of methane, oxygen and air (11).

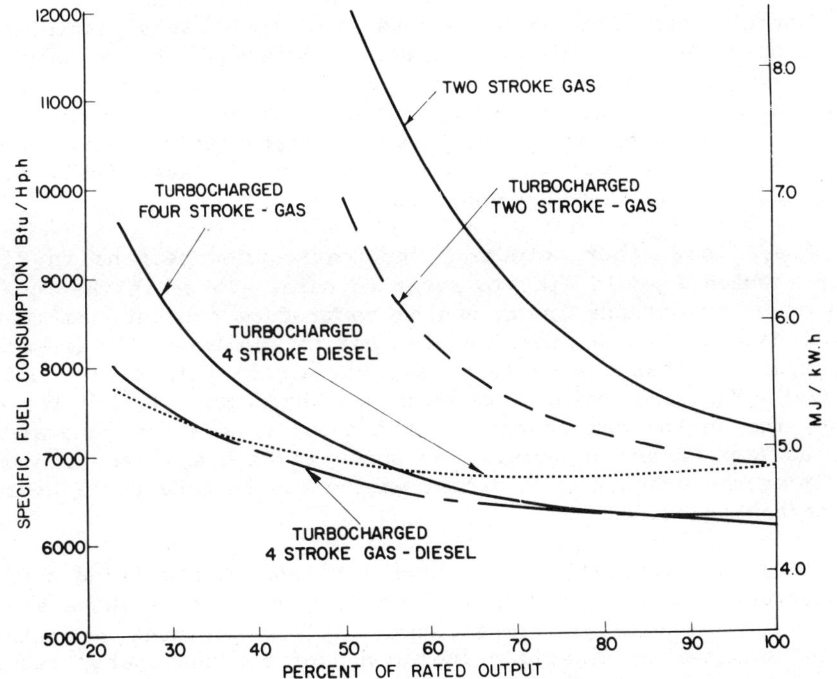

FIGURE 13. Typical variation of the specific fuel consumption with load of spark ignition and diesel engines fueled with methane (12).

of the methane added will be fully oxidized. Of course, at such small total equivalence ratio the amount of gaseous fuel added is small and hence the unreacted fuel will be equally small. In any case there are ways of improving upon this tendency; for example, through the use of larger pilots or by encouraging the heating of the intake charge through preheating or exhaust re-circulation. Moreover, partial throttling of the intake air if need be can be used, to render the mixture slightly richer.

One important factor that needs to be considered is how big a liquid pilot is to be used. Most of the authors in this book indicate that extremely large pilots were being used. So much so that it can be described as diesel operation being topped up with some gaseous fuel. This is, in my view, perhaps unnecessarily restrictive and potentially costly. The amount of diesel fuel being injected can be cut down to extremely small amounts. In fact, some time ago we ran a research engine on methane with a pilot quantity as low as 3 to 5% of full diesel operation. Of course, it is not necessarily being advocated that this is the sort of level to operate at, but at least it indicates the reduction in diesel quantity that can be achieved and hence the potential economy involved.

Normal operation with methane can yield very satisfactory operation with most diesel engines. However, it is only when very high power outputs or very high intake temperatures are involved that the problem of knock, even with the very knock resistant fuel methane, is encountered. Fortunately, the knocking region is out of most of common operations unless highly supercharged diesel engines are employed.

There is another potential application for methane in diesel engines which I would like to refer to here. There is the possibility of using methane in an engine to produce exhaust gas that is useful chemically. A diesel engine can be run carefully on methane and oxygen so that the exhaust gas will almost entirely be made up of synthetic gas, carbon monoxide and hydrogen (11). In other words, the engine can operate simultaneously as a gas generator as well as a producer of power. As shown in Figure 12 as much as 85 to 90% of the exhaust of a diesel engine can be made up of hydrogen and carbon monoxide.

In conclusion, it will be useful to show typically a comparison between a spark ignited gas engine operating on methane alone relative to a similar diesel engine also operating on methane. Figure 13 shows the specific fuel consumption: the energy required to produce a unit amount of work as a percent of rated engine output. For an engine operating as a spark ignition gas engine, the two-stroke mode is of relatively poor efficiency; while turbocharging the two-stroke, particularly when it is well done, can improve the specific fuel consumption significantly (12). The four-stroke turbocharged engine operating on gas and spark ignited, has an efficiency that is superior to that of the two-stroke. On the other hand, diesel-gas operation is generally superior to that of spark ignition over the whole load range. Turbocharging the gas-diesel will produce a relatively lower efficiency at light load than the corresponding diesel. However, over the bulk of the higher output range, the turbocharged gas-diesel is far superior to the straight turbocharged diesel both in terms of efficiency and power output. Therefore, the turbocharged gas-diesel is much superior to the spark ignition engine or even the corresponding straight diesel.

Figure 14 shows the mean effective pressure, in other words the power output, against the total mixture strength (i.e. gas + diesel) being fed to the engine for different diesel pilot quantities over the light load range (7). A relatively large pilot will consume almost all the fuel being supplied and the resulting power output is directly related to the fuel quantity. If a very small

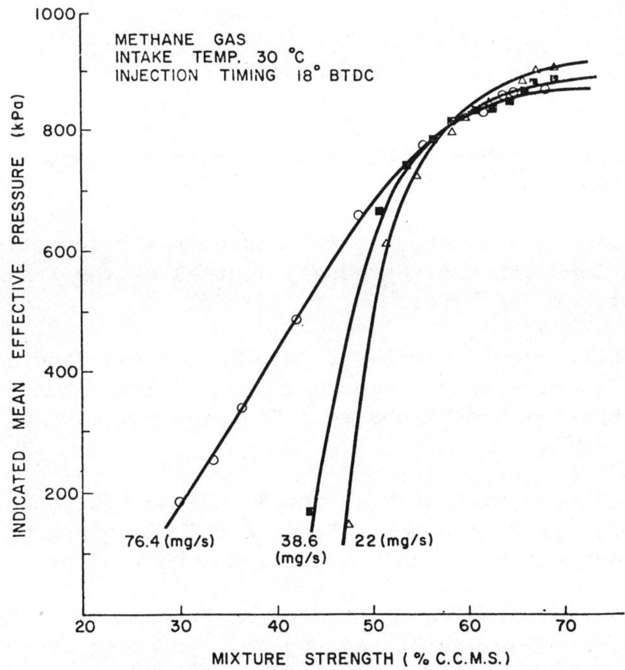

FIGURE 14. Indicated mean effective pressure variation with mixture strength of a dual fuel engine fueled with methane for three different values of pilot quantity at constant injection timing (7).

pilot were to be used, then some of the fuel will not be fully utilized initially, resulting in lower power output. However, at much higher loads there is no longer a need to use large pilots and there appears to be a definite improvement associated with using a very small pilot. This would clearly point out the desirability of having a variable diesel pilot to methane ratio over the whole load range. At light load a relatively large pilot is used, but at higher loads the pilot quantity can be reduced yet still have very good performances. Controlling the pilot to methane ratio to an optimum level is potentially a job that is highly amenable to automatic control and the use of mini-computers. This is a task that is waiting for effective development.

In conclusion, the diesel engine appears to be extremely attractive for the efficient utilization of the gaseous fuel methane; perhaps it is even more attractive in this regard than the spark ignition engine.

REFERENCES

1. Boyer, R.L., "Status of Dual-Fuel Engine Development", S.A.E. Journal, May 1949.

2. Karim, G.A. and Singh, R., "A Thermodynamic Investigation into the Combustion of Methane", Journal of the Institute of Fuel, October 1967, Vol. 40, p. 447-455.

3. Khanna, S.L., "An Experimental and Analytical Study of the Combustion Kinetics of Methane Employing a Flat Flame Burner",PhD Thesis Department of Mechanical Engineering, The University of Calgary, 1972.

4. Karim, G.A. and Klat, S.R., "The Knock and Autoignition Characteristics of Some Gaseous Fuels and Their Mixtures",Journal of the Institutue of Fuel, March 1966, Vol. 39,p. 109-119.

5. Karim, G.A., "A Review of Combustion Processes in the Dual Fuel Engine - The Gas Diesel Engine", Progress in Energy and Combustion Science, 1980, Vol. 6, p. 277-285.

6. Sulzer Brothers Co., Winterthur, Switzerland, "Dual Fuel Engines for Ship Propulsion Machinery".

7. Karim, G.A., Klat, S.R. and Moore, N.P.W., "Knock in Dual-Fuel Engines", Proceedings of the Institution ofMechanical Engineers, March 1967, Vol. 181, p. 453-466.

8. Karim, G.A. and Burn, K.S., "The Combustion of Gaseous Fuels in a Dual Fuel Engine of the Compression Ignition Type with Particular Reference to Cold Intake Temperature Conditions". Proceedings of SAE Congress, February 1980, Paper No. 800263.

9. Land, M.L., "Performance and Operation of Spark Ignited Gas Engines", Institute of Mechanical Engineering 1967, Vol. 181, Part I.

10. Karim, G.A. and Khanna, S., "The Effect of Very Low Air Intake Temperature on the Performance and Exhaust Emission Characteristics of a Diesel Engine", Proceedings of SAE National Powerplant and FCIM Meeting, September 1974, Paper No. 740718.

11. Karim, G.A. and Moore, N.P.W., "The Production of Synthesis Gas and Power in a Compression Ignition Engine", Journal of the Institute of Fuel, March 1963, Vol. 36, p. 98-105.

12. Segaser, C.L., "Internal Combustion Piston Engines", Rep.No. ANL/CES/TE 77-1, National Technical Information Service,U.S. Dept. of Commerce.

13. Karim, G.A. and Ali, A.I., "Combustion, Knock and Emission Characteristics of a Natural Gas Fuelled Spark Ignition Engine with Particular Reference to Low Intake Temperature Conditions", Institution of Mechanical Engineers, August 1975, Vol. 189 24/75.

THE HOME-RECHARGEABLE

NATURAL GAS COMMUTER CAR

Robert T. Axworthy
Vice President, Marketing
Corken International Corporation

Oklahoma City, Oklahoma

In the United States and Canada there has been a great deal of attention given to the use of gaseous fuels for fleet vehicles, but there has been little concern given to the general public's use of compressed natural gas in cars. This chapter represents a modest proposal for such a system.

Almost 12 years ago our company developed the concept of a home-rechargeable natural gas commuter car. We built such a vehicle for a cross-country race sponsored by MIT and Cal Tech. They called it the great Clean Air Car Race of 1970. The car, which we called "The Gasser", was driven 3600 miles in the race from Boston, Massachusetts to Pasadena, California at an average speed of about 70 mph on the equivalent of $7.40 worth of natural gas at the prevailing Oklahoma City retail price. After the race, I used the vehicle as a personal car and operated it for an additional 70,000 miles.

The original concept was to develop a vehicle to compete with the rechargeable electric car. In my state of Oklahoma, a high percentage of our electricity is generated with natural gas. It seemed a bit ludicrous and certainly a profligate waste of energy to burn natural gas to generate electricity, with two-thirds of the energy wasted in the process, transmit and transform this power several times, meter it into my home and then use it to charge a rather inefficient battery, to drive a vehicle which might better be driven by natural gas in the first place!

At this point I'd like all of you to wipe out your preconceived notions of the design parameters of a CNG commuter vehicle. We should not assume that a viable, mass production gas car will have the same characteristics of capacity, range and pressure as the converted gasoline cars with which all of us are so familiar. But let's start with a blank sheet of paper and follow with me in the logic of designing a concept from scratch.

Very early we rejected the idea of public fueling stations. For any major impact on the U.S. transportation market, public stations would represent a massive capital outlay for even a minimal nation-wide CNG fueling program. As can be learned from the New Zealand experience, the minimal "On-Line" filling station represents a capital outlay in excess of $250,000. In the U.S., if we were to have only the number of fueling points for CNG as we now have for LP gas - about 20,000 - we're looking at a capital outlay of more than $5 billion! And, as in New Zealand, the chicken and egg syndrome would apply. Basically, the fueling points would have to be installed before the vehicles could be sold.

In selecting the design parameters for our home rechargeable natural gas commuter car, it soon became apparent that "range" and "recharging time" were the two elements without answers. No range was long enough and no charging time was short enough! Our investigation showed that most cars go fewer than 50 miles per day. So a 50 mile vehicle should be valid. But not so! The problem with the 50 mile car occurs on the 51st mile ...you're dead in the road and need a tow! A 75 mile or 100 mile vehicle would have the same problem. As is usual with all mechanical equipment, it must be designed for the exception, not the rule!

And this is the point at which the natural gas vehicle design invariably bogs down. In an attempt to give the car the range to go anywhere, larger and larger tanks must be added and higher and higher pressures must be used. This increases both vehicle weight and cost and reduces the economic advantage of the gas vehicle. Similarly, continued escalation of tank pressures increases charging equipment costs and increases the energy required to compress gas to higher and higher pressures.

And so we arrived at the basic key to a viable home rechargeable vehicle.... the necessity for a range-extending technique. (The electric vehicle people have recognized this problem with the so-called Hybrid electric vehicle.) And yet it had to be compatible with the entire gas vehicle concept. Fortunately, the natural gas commuter car has a completely compatible standby or reserve fuel available in LP gas. Thus a five or six gallon LP gas reserve allows for emergency use to extend the commuter vehicle range and an additional 150 to 200 miles and is easily refillable for use during extended trips. With over 20,000 LP gas fueling points all over America, fuel availability presents few problems.

Fuel Availability

Any viable home recharged vehicle must be based on readily available energy. With over 43 million homes in America piped with natural gas, this requirement is certainly met.

The viability of the home-recharged compressed natural gas commuter car can be dramatically summarized with these facts: in the United States, last year we used about 20 trillion cubic feet of natural gas. Some of my associates in Oklahoma tell me that our Anadarko Basin deep gas structure alone possibly holds over 100 trillion cubic feet of gas. And one trillion cubic feet of natural gas will fuel 20 million commuter cars for a year!

And while LP gas supplies are not sufficient to fuel any major proportion of the nation's vehicles, there is plenty of LP gas available for commuter car standby or backup purposes.

Vehicle Selection - The Commuter Car

But first let's look at the vehicle itself. Commuters are the key to our present energy problem. With inadequate and obsolete transit systems, most Americans must use automobiles to get to work. And the classic American family car is a rather inefficient personal transportation device. While the group attending the conference is knowledgeable in the field of vehicular transportation, very few of you have a good feel for the overall efficiency of automobiles when they are used to move people. I estimate that my 1977 vintage family car wastes 99 and 44/100% of the fuel I use to drive to work. Believe it? Well, 70% is lost in the basic inefficiencies of the internal combustion engine as we are all aware and another 10% is lost to vehicular friction in various guises. The remaining 20% doing usable work is moving a 4,000 pound car and a 150 pound payload ... me! That 25 to 1 ratio of dead weight to payload brings our overall transportation efficiency to less that 1%! Not really anything to brag about in a period of energy shortages. And all this points up the fact that the commuter car should be small - an honest two passenger car with a curb weight well under 2,000 pounds. A commuter car is one of the two elements that are missing to make the home rechargeable natural gas car viable.

The Gas Charging Device

The other missing element of our system is the device to transfer the natural gas from the home gas system into the fuel storage tanks on the vehicle. The gas must not only be transferred, it must be "transformed" from a pressure of a few inches of water up to many hundreds of pounds per square inch pressure.

And this is where my company and I fit into the equation. As a manufacturer of small gas compressors, Corken has designed a home-fueling package to recharge the vehicle tanks. We ultimately settled on design pressures and capacities which we felt to be optimum for this service. Our first thought was to use design pressures that had been used successfully for fleet motor fueling ... about 2500 psi. Our studies showed us, however, that the efficiencies of very small compressors were such that the value of the electricity driving the compressor motor could become a major fraction of the value of the gas being compressed! Also, when pressure is doubled, the motor size must almost be doubled for the same capacity. A key design parameter of the home fueling compressor was that it must be capable of being powered by 110 volt power available in the home. This meant quiet, fractional horsepower motors.

We found that lighter, lower working pressure tanks were available that could give us a tank weight to capacity ratio that was similar to the higher pressure tanks. However, the tanks would take more space for a given capacity.

Our optimum pressure selection was 1000 psi. This pressure could be reached with a fractional horsepower motor for an electric cost of less than 7¢ per hundred cubic feet of compressed gas. We used a standard DOT 3AA 1000 cylinder weighing about 30 pounds, approximately 30 inches by 9 inches in diameter, and manifolded four tanks in our vehicle. This was a total weight of about 120 pounds which gave us a vehicle range of around 75 miles for the small two passenger Opel GT vehicle which we converted. For comparison, electric batteries for similar range on our vehicle would have weighed in excess of 1200 pounds. Also, for comparison, the tanks we bought 11 years ago are now worth more than we paid for them. If I had gone electric, I would now be on my third set of batteries at a cost of over $2000 per battery set.

Our backup fuel supply of LP gas in "The Gasser" is contained in a 10 gallon tank which we used because it fitted in the space made available when the gasoline tank was removed. It is twice as large as would be necessary for the average commuter car. But this extended our range on the open highway by an additional 300 miles.

The Corken charging compressor was designed to be as service free as possible. We used "dry cylinder" construction with no oil lubricant in the cylinders. This not only eliminated the possibility of putting oil contamination into the storage system, but it meant that oil would not have to be continually added to the compressor crankcase. For the 65 compression ratios involved, a three stage compressor design was selected.

The Charging Process

When the vehicle returns home, it is immediately connected to the charging device. The compressor is turned on and it continues to operate until the pressure switch shuts it down at 1000 psi. If the tank starts out completely empty, the charging process takes about 2 1/2 hours for the 240 SCF capacity. The fueling plug disables the ignition and the car can't be started until the filler hose is removed. The compressor is located outside the garage and fueling is done in the open air.

Fringe Benefits From The Use of Two Fuels

In addition to the range extending feature of the backup LP gas fuel, there are a number of additional benefits. In a period of recurring energy crises, it is inevitable that there will be periods and areas of product shortages. With two fuels, the car owner's options are increased tremendously. Last winter in Boston for example, natural gas limitations would have increased LP gas usage in commuter cars.

The availability of the standby fuel system also minimizes the importance of a "fast charge". Conventional rechargeable vehicles require super fast charges for those special situations which invariably will confront a car owner. And again, the LP gas backup fuel makes this requirement unimportant and the charging device can be selected for optimum performance and minimum size.

Why Not Gasoline as the Backup Fuel?

One of the basic advantages of the gaseous fuels is their inherent high octane value. This would allow for the use of higher compression engines with improved fuel efficiency and performance. And the inherently clean combustion products of the gaseous fuels would allow for much more freedom in the design of emission controls. It would be unthinkable to burden the engine design with the compromises needed to get optimum performance out of both the gaseous fuel and gasoline, and it is not necessary! The range extending fuel is not normally used. It is not so much an alternative fuel as it is a backup.

While no two fuels would have the same optimum compression ratio and ignition timing requirements, the compromises between propane and natural gas are fewer and less traumatic to the engine design than those required between natural gas and gasoline. This concept would result in a true gaseous fuel engine design wherein most of the efficiency stealing emission controls would be minimized or eliminated.

Economics of the Home-Rechargeable Natural Gas Commuter Car

In Oklahoma City today, our natural gas motor fuel costs me about 45¢ per an equivalent gallon of gas. This includes the electrical cost of compression and our state road tax. With gasoline costing me about $1.30 per gallon, I save about $.85 per equivalent gallon. This affords considerable fuel savings that could pay off any premium in cost the gas vehicle might have.

In the very small production quantities we are now looking at for the home charging compressor package, a price in the $1500 range is anticipated. For large scale production, the home charger should sell for around $500. The gas vehicle premium should be modest, but very large-scale production would be the key to holding these costs down.

Safety of Gaseous Fuels

There has been a great deal of work done on safety studies of the hundreds of thousands of vehicles presently in operation all over the world on both compressed natural gas and liquified petroleum gas. A number of very comprehensive experimental crash tests have been made.

I will not belabor this presentation with a rehash of these results. Suffice to say that, even with "leading edge" technology in many instances, the safety record of gaseous fuels has been exemplary. Primarily, this has been the result of two factors: first is the use of high standards and "approved" equipment for gaseous fuel systems and second is the inherently strong fuel storage tanks used with the compressed gases. While classic petroleum vehicular fuels are stored on board in thin sheet metal containers which are extremely vulnerable to impact damage, both LP gas and CNG tanks are sturdy vessels made from high strength materials extremely resistant to collision damage.

Fuel Distribution

A basic consideration in the home-fueled gas or electric vehicle is the inherent efficiency of fuel distribution. With natural gas distribution systems, meters and billing systems already in place, motor fuel almost gets a "free ride". And since it is a 12 month steady load, it must be considered as a preferred type of business to the gas utility. Other alternate motor fuels would be hard pressed to approach this distribution efficiency. Also, with most alternate fuels, there would be considerable delay in implementation and tremendous capital costs in the development of the fuel handling distribution infrastructure. With natural gas in the United States, this is already in place. With the vehicles and chargers in place, we could start fueling 20,000,000 vehicles in the United States immediately without appreciable impact on our gas distribution network, during most of the year.

Conclusion

I have great concern over the decline in the U.S. national sense of urgency in coming to grips with the imported oil problem. The current international oil glut is illusory and certainly has no long-term significance. The fact is, that when we built "The Gasser", the United States imported $3 billion worth of oil. Last year this was $80 billion!

We in the United States must take immediate action to utilize domestic resources and I submit that the natural gas commuter car is the most effective single action which we can take to control our international trade deficit. With hundreds of thousands of vehicles already operating on natural gas, and far more than this number of LP gas vehicles in everyday operation all over the world, it appears non-productive for us to belabor the viability of gaseous fuels per se. The economics, safety and availability of this fuel have been amply documented. It is time to move on to the logical next step. That is, of course, the manufacture of vehicles specifically designed for gaseous fuels.

Even in the most viable of the proposed non-petroleum vehicular fuels, there exists a number of basic inertial obstacles to their use that clearly limit any early impact on vehicular energy problems. Technology must be developed, tested and refined. Massive production facilities must be planned, financed and built. A distribution infrastructure for the transportation, storage and delivery to users must be set up. While this does not preclude the use of many alternate fuels, it does create an initial massive outlay of capital if any major impact on transportation fuel usage patterns is to begin.

The home-refueled compressed natural gas commuter vehicle is unique in that the production and distribution system for this fuel is already in place. The vehicle and charger are the only missing elements. It would not be an exaggeration to state that, given these two missing elements, tomorrow morning 20 million U.S. commuter cars could be fueled with natural gas with no greater impact than a smug smile on the faces of a large number of gas utility executives! Forty-three million homes are piped and metered for natural gas. Except for peak usage periods, the gas distribution system in America is not operating at its capacity.

The electric utility people will be smiling too since they

will be getting a nice off-peak load, for most of these natural gas vehicles will be charged up at night utilizing an electric motor to drive the charging compressor.

In the final analysis, no experts are omniscient enough to tell us the future relative availability of domestic oil and gas. But the very clear fact remains that no matter what the source of vehicular fuels will be, whether from conventional wells or from other sources, natural gas will have an advantage in both energy efficiency, distribution efficiency and economics over the liquid fuels. Processing natural gas from coal, for instance, typically requires about half the energy required to process the same coal into a liquid fuel. Thus, it would seem prudent to develop a new breed of gaseous fueled vehicles which could operate directly on natural gas without the need of further energy losses in processing it into liquid fuels. The distribution infrastructure is in place. Conventional automotive technology is all that is required. The only missing link is the vehicle... a mass-produced commuter car designed specifically to use the unique advantages of the gaseous fuels... high octane for high compression engines and inherently clean burning for minimal emission control equipment. With the fuel, the technology and the incentive, what on earth are we waiting for?

COMMERCIAL EXPERIENCE WITH ALTERNATIVE FUEL PROGRAMS

A RETAIL MARKETING STRATEGY FOR METHANE

W.C. Dunning

Caltex Oil Limited

New Zealand

Methane may very well be the fuel for the future. I say "may" because we can sit here and talk about the production, the processing, the conversions but until we go out and market the product and sell compressed natural gas (CNG) to the customer, methane will not be the fuel for the future. I will discuss Caltex's plan for marketing CNG in New Zealand.

Caltex, incidentally, operates only in the eastern hemisphere but we do have associated companies in the western hemisphere through our shareholders: Standard Oil of California and Texaco Incorporated.

Why CNG in New Zealand?

New Zealand is a fertile farming country with an area of 26.9 million hectares and is similar in size to the British Isles and Japan. Relative to its size, however, its population is small, at only 3.2 million.

New Zealand is a trading nation and one of the largest exporters in the world of butter, cheese, wool and meat. In the last 10 years, New Zealand has also become a significant exporter of timber and wood pulp. Despite its enviable export role, the rising prices of imported petroleum products have contributed to a worsening of the balance of payments. The national oil bill has risen from $114 million in 1973 to an estimated $1.6 billion in 1981.

Although New Zealand is rich in natural energy resources - hydro electricity, geothermal power, natural gas and coal - it is

85% dependent on imported oil and products for transportation fuels.

In 1979 the Ministry of Energy announced a comprehensive plan to develop its own energy resources to lessen its dependence on imported oil products and to improve its trading balance of payments.

The cornerstone of the Government's energy plan is the utilization of its huge reserves of natural gas.

The Maui and Kapuni gas fields in the North Island of New Zealand have reserves in excess of 5.4 trillion standard cubic feet of gas, with Maui having a projected life in excess of 30 years. Through 50% ownership of the fields with a consortium of private companies, the Government signed an agreement to purchase all the natural gas from the joint venture. The Government has established the Natural Gas Corporation (NGC) to market the gas and is thereby the sole primary supplier of natural gas.

To maximize the use of this resource to achieve its energy goals, Government is giving top priority in three new areas of natural gas utilization:

- 16% for synthetic gasoline production.
- 9% for methanol production for export purposes to obtain additional foreign exchange.
- 10% for CNG and LPG production.

We are primarily concerned with the direct use of gas as a transportation fuel. Central to this element of the Energy Plan is the Government's goal of converting 150,000 vehicles to the use of compressed natural gas as a transportation fuel by the end of 1985.

It is estimated that the CNG utilized in the 150,000 vehicles converted will replace 16% of the gasoline market of 38,000 barrels per calendar day (BPCD) or 6,000 BPCD. Upon completion of the program, savings in foreign exchange to the country through this substitution will be in excess of $100 million per annum at today's prices.

Additionally, the increased production of natural gas for CNG will permit the annual recovery of almost 1 million barrels of condensate for refining at the New Zealand Refinery Company's plant. This condensate will substitute for imported crudes at an annual foreign exchange savings of $44.5 million.

Government's Role and Incentives

Central to the success of the program in New Zealand has been the attitude and role of Government. As previously stated, the conversion of 150,000 vehicles to CNG by the end of 1985 is an integral and vital part of their energy policy.

Through agreement with the producers' consortium the Government is the sole purchaser of natural gas and therefore controls the basic supply and selling price of natural gas to users such as private and public natural gas companies.

Additionally, as in many countries, prices of primary fuels are Government controlled in the market place.

Putting these two control elements together, i.e. cost of raw material and final product selling price, the Government has the unique opportunity of offering potential CNG users a desirable cost advantage as an incentive for conversion. The Government has made a public commitment to maintain a substantial margin between the retail price of petrol and CNG and not to apply taxation to the disadvantage of CNG. Currently CNG costs the consumer approximately 50% of the equivalent petrol retail price.

While CNG pricing is unquestionably the most important incentive for consumers to convert, the Government also recognized the relatively high cost of consumer conversions and the high costs required to establish CNG dispensing outlets. To assist in these initial expenses, the Government has established a generous incentive program aimed at accomplishing its conversion goals:

- A $200 grant to the motorist on each CNG conversion kit.
- First year 100% tax write-off of the cost of converting a business vehicle to CNG.
- Extension of the tax write-off to cover kits fitted on new vehicles on assembly lines.
- A "development" grant of 25% of the capital cost of providing CNG as a motor fuel. With the exception of land and buildings, this grant is applicable to most of the capital cost of establishing filling stations.
- The balance of the "development" expenditure qualifying for the 25% grant is permitted to be written off 100%, in the first year, for tax purposes.
- Companies with a low tax incidence cannot take advantage of the first year accelerated depreciation. Accordingly, with effect from April 1, 1982, the scheme will be broadened by replacing the accelerated tax write-off with Government loans for up to $500,000 per project.

The savings from these government incentives, coupled with the price differential between the two fuels, offer a quick payout for conversion costs.

For a typical business user travelling 16,000 miles per year, please note the effects of incentives in the following simplified analysis:

	$
Annual Gasoline Cost	1,969
Annual CNG Cost	1,052
Fuel Savings	917
Modest Maintenance Savings	100
Total Savings Before Tax	1,017
Total Savings After Tax	559
Cost of Vehicle Conversion	1,620
Less Government Grant	200
Tax Shield	640
After Tax Investment	780
Payout	1.4 years

Furthermore, the Government has instituted a major community and public relations program utilizing the media, to promote the conversion and use of CNG, and has been the major coordinator of sellers' and users' organizations, leading to greater cooperation and more effective operation of these organizations. They have taken the lead in converting Government vehicles and encouraged the use of private outlets for these vehicles.

While Government's role will vary from country to country depending on the special circumstances in each country, in today's world they can play a major and key role in the success of any CNG project.

Caltex and CNG

Recognizing the importance of CNG to the national interest of the country, the planning department of my company initiated a preliminary study to ascertain the feasibility and viability of our entry into CNG marketing.

Early consumer research indicated that conversions to CNG by fleet operators and private motorists were well below the Government's goal because of:

- A lack of refueling outlets in key areas.
- High costs of conversion.
- A lack of driver understanding about the "drivability" of a CNG powered vehicle.
- Concerns relative to safety of CNG as a transportation fuel.
- A reluctance to "change".
- A lack of a concerted selling campaign aimed at fleet users.

As the New Zealand Government has established an excellent incentive program for both potential users and suppliers, it appeared that the major obstacles should be easy to overcome.

Further market research, however, indicated to Caltex that the CNG conversion program was faced with a "chicken and egg" situation. With the price advantage of CNG and government incentives, potential users wanted to convert but were waiting for refueling outlets to be established to assure their continuous supply. On the other hand, potential CNG resellers looked at the high investment cost required to establish a CNG dispensing outlet and were reluctant to make such investment until a market was more firmly established.

Caltex's economic analysis indicated that our entry into the CNG market was a viable commercial proposition, assuming the achievement of the Government's conversion objectives. For these objectives to be achieved, the "chicken and egg" attitude had to be reconciled.

Our decision was that the customer was paramount and that the seller has the obligation to make the product available to the consumer. Accordingly, we devised a program to establish a chain of more than 50 high volume CNG dispensing outlets in the key centers of the North Island of New Zealand. These outlets are to be developed through revamping present petrol outlets: they will be dual fueled and offer full customer service as does a regular petrol reselling outlet. Service capability of outlets is to be 300 to 500 cars each day.

Quality and Benefits as a Motor Fuel

As previously stated, potential consumers lacked understanding of the "drivability" of CNG and its benefits as a motor fuel.

In developing a CNG marketing program, you should first assure yourself of the quality and composition of the natural gas available and its advantages and disadvantages versus competitive fuels such as petrol and LPG.

The important quality factor to the end user is the calorific value of the gas as determined by the number of British Thermal Units (Btu's) per standard cubic feet (SCFT). As you know, the major elements of typical natural gas are methane and carbon dioxide. While lower in percentage of composition, ethane, propane and butane are extremely important components of natural gas in maintaining higher Btu's. The excessive stripping for other use of these elements from the natural gas may make CNG uncompetitive versus petrol and LPG.

Available gas in New Zealand today has a Btu value of 1140 per SCFT from the Kapuni field and 1023 per SCFT from the Maui field. Comparing Btu's with other automotive fuels is difficult because CNG is measured as a gas and petrol and LPG are measured as liquids.

By comparison, in straight calorific terms, 0.81 cubic meters of gas roughly equate to one liter of petrol. Road tests in New Zealand, however, indicate that 0.72 cubic meters of gas equate to one liter of petrol. The actual increased CNG efficiency achieved in automotive use results primarily from the higher octane rating of CNG, i.e. 130 RON (research octane numbers) versus 96 RON for petrol. Based on actual performance and the value of New Zealand gas, CNG on a per liter equivalent has a calorific value equal to 89% of petrol.

Comparisons with competitive fuels will give you further confidence in the marketability of the product and lay the groundwork for your advertising, sales promotion and community relations programs.

The following are some examples of our findings in New Zealand:

1. Advantages compared to petrol (excluding safety benefits):

   (a) Lower running costs through reducing oil and spark plug changes. Lubricating oil life extended since CNG does not contaminate and dilute oil. Because of the absence of lead, lead fouling is completely eliminated.
   (b) Reduced engine wear and therefore reduced maintenance. CNG does not have washing effect of petrol so the lubricating oil in the area of the cylinder bore, pistons and rings can be more effective.
   (c) Easy starting and increased engine efficiency. CNG provides excellent cold starting and cold engine performance as it enters the cylinders in a vaporized form. It is also clean burning. Exhaust emissions are approximately 90% lower and virtually no carbon is formed, reducing deposits in the valve and ringbelt areas.

(d) Smooth engine performance. Petrol and air mix in an engine tend to separate out. CNG mixes readily with air and does not separate out, resulting in more even cylinder distribution.

(e) As previously mentioned, reduced exhaust emissions make CNG a highly acceptable automotive fuel from an environmental point of view.

2. Disadvantages compared to petrol:

(a) Restricted vehicle range. Because of its relatively low "energy density" the distance that can be travelled on one CNG cylinder (usually called a tube) is typically 120 to 150 km compared to 400 to 500 km on petrol. Even with two storage tubes you could only increase distance travelled to 240 to 300 km.

(b) Power loss and reduced acceleration. Compared to petrol, CNG users commonly encounter a 15% reduction in power. Drivers quickly adjust to this power loss and with present world-wide speed limitations most drivers are unaware of any power loss.

(c) Weight and bulk of storage tubes. CNG tubes take up truck or payload space and add to vehicle weight. Smaller sedan-type vehicles can be comfortably fitted with up to two nine cubic meter tubes, while small service vans (2 to 4 tons) can be fitted with three or four larger 20 cubic meter tubes. Looking at the sedan, the two tubes filled with CNG would weigh approximately 250 lbs. versus the filled petrol tank weighing 100 lbs.

Safety

As CNG as an automotive fuel is relatively unknown, the next question you and your potential customer would want answered is: "Is CNG a safe product?"

While all automotive fuels are safe when handled, transported and dispensed with care, there are many public misconceptions regarding gas. Largely this is because the true facts are unknown to the public.

Caltex's basic thrust is that the natural gas dispensed into your vehicle is the same gas the consumer has been using safely for years for home cooking and heating, compressed by necessity for dispensing, storage and use in a vehicle.

All aspects of the safety on CNG are considered to be better than those associated with competitive fuels:

(a) Lighter than air (approximately 0.55 the weight of air) allowing for rapid dissipation if leakage occurs. CNG will not "puddle" as does LPG or petrol.
(b) CNG requires a high ignition temperature in air, approximately 704°C versus 455°C for petrol.
(c) While it can be asphyxiating, it is non-toxic.
(d) Limited explosion range - 4 to 15% as a percentage by volume, either side of which it will not burn.
(e) Usually odorised to detect leaks easily should they occur.
(f) CNG conversion equipment and the installations of equipment and tubes into a vehicle, are governed by strict government regulations, including certification of installers.

Facts on the safety aspect of CNG play an important role in the marketability of the product.

Supply, Costs and Terms

As in marketing any product, an assurance of raw material availability or security of supply is vital. The cost must be clearly indicated in any agreement, as well as elements related to cost, such as credit terms and escalators. With the level of investment required in a CNG outlet, short-term supply contracts must be avoided unless the agreement includes viable option periods.

In New Zealand there are a number of branch pipelines from the gasfields to key urban locations and major commercial users. In the main urban areas gas is sold by the Government-owned Natural Gas Corporation (NGC) to local gas companies who in turn sell to consumers.

In obtaining the required security, Caltex first approached NGC. the primary supplier, and obtained written assurance that quantities required for our CNG program would be made available to our direct suppliers. This basic assurance also permits your direct supplier to negotiate with confidence your supply requirements.

Of major importance in the supply of natural gas is the pressure delivery rate in pounds per square inch (psi). Pressure available in New Zealand from gas companies is normally at 5 to 115 psi's although at some locations in Wellington and Auckland 50 psi is available. As delivery pressure must be compressed to 3600 psi the lower the psi supplied the more capital investment required in compressors.

As a consequence, it is important that you should negotiate

pressure guarantees or commitments for each retail outlet or compressor location.

The duration of the contract will vary at different locations; however we found that the direct supplier was willing to offer the same period of supply as he had obtained from the prime supplier, including option periods.

In a major program such as ours in New Zealand, the high volume required guarantees that the price from the supplier is the best obtainable from the rate schedule. In some cases we negotiated on the basis of the total volume utilized within the supplying area rather than on a per outlet basis and were able to achieve further percentage discounts off the most favourable rate.

Obviously, the supply price is an important factor in a program such as ours as it is necessary to obtain revenue required to cover a margin for our resellers, cover all costs and be assured of a reasonable rate of return on our investment.

Favourable credit terms from your supplier can be of major importance to a retail CNG program.

In this day and age of fluctuating energy prices, any long term supply agreement will include escalation clauses. As the basic reason for escalation clauses is to protect the supplier from increased costs from the prime supplier you should negotiate equivalent terms in your agreement with your direct supplier. Equally important, your agreement with the reseller must include the same terms and conditions.

Consumption Estimates and Market Share

Concurrent with our investigation of other areas of market research and basic to our management decision was a study of consumption.

As there was essentially no established market for CNG, consumption estimates had to be developed on the basis of potential users, and the ability of those concerned, i.e. government and private sector, to bring about required vehicle conversions.

We first assured ourselves of the Government's determination to achieve its goal of 150,000 cars converted to CNG use by the end of 1985.

In reviewing the vehicle potential, we saw that New Zealand has in excess of 1.5 million registered automobiles and goods service vehicles of which 750,000 are in areas serviced by the natural gas pipeline network. An additional 60,000 vehicles are in areas that are serviceable by tube trailers. These latter, in addition to constituting growth areas, are also resort and service areas closely related to the urban cities of Auckland, Wellington and Hamilton and therefore important in completing the desired network of CNG dispensing outlets.

As Government conversion incentives are largely aimed at commercial users, we ascertained that this group represented 16% of the vehicles in New Zealand. A further 10% are considered business related, i.e. sales representatives, company plan, etc. In this regard it is important to note that in New Zealand "company plan" cars down to middle management level is standard practice. As our target area was primarily the large urban areas, we could expect the percentage of business registrations to be in excess of 26% and would consume a disproportionate amount of total consumption.

Estimating a realistic rate of conversions through 1985, Caltex's estimate of 142,000 vehicles potential was close to the Government's target of 150,000.

Additional assumptions included:

(a) An annual mileage of fleet vehicles of 16,000 miles (c.f. national average is between 9,000 and 10,000 miles per annum).
(b) Final consumption rate of 18.2 miles per US gallon.
(c) Cars will be dual fueled and in early years will operate on 70% CNG and 30% petrol (Southern California Gas Company estimate).

We estimated that CNG would replace 16% of the existing petrol market of 38,000 BPCD or 6,000 BPCD. This equates to approximately 30% of the petrol market in the areas to be serviced by CNG.

CNG demand would be enhanced above these estimates, if mandated demand restraints were instituted (i.e. petrol rationing, carless days, etc.), if Government instituted additional incentives and if strategically located outlets permit dual fueled vehicles to operate entirely on CNG.

Based primarily on four major competitors in the market, our market percent in liquid fuels, the strategic location of our chain and the early penetration into the CNG market Caltex estimated its long-term share of the market, with a higher percentage in the early years.

Methods of Distribution

Important to the development of any CNG project is an analysis of methods of distribution. Early in the development of the project, Caltex decided that CNG would be marketed through its existing outlets as opposed to establishing separate grass-roots outlets dedicated to CNG.

In New Zealand, Caltex has elected two separate methods of distributing and marketing CNG.

The first method is to install 40 to 80 HP compressors (as required) at retail outlets connected directly into the local gas company's existing city pipeline reticulation system. We call these outlets "On-Line" stations.

The second method is to establish a major compression station ("Mother" station) utilizing approximately 500 HP compressors sited adjacent to the main transmission pipeline to feed the compressed gas direct into trailer tube storage for road delivery to a retail outlet ("Daughter" station) for connection to the outlet's vehicle dispensing system.

While the "On-Line" method has certain advantages such as minimum investment cost and minimum space required at the retail outlets, it could not be used as a total concept due to limitations of existing gas pipeline systems and pressure capacity availability.

At the "On-Line" site gas is received at the gas company's regular pipeline at between 5 and 50 psi. As it passes through a gas meter Caltex takes title to the gas and simultaneously resells the gas to the dealer. The gas is then compressed at a delivery pressure up to 3600 psi. The compressed gas is stored in storage cylinders sized to accommodate peak demands. The gas is dispensed into the vehicle at a maximum pressure of 2400 psi from the storage cylinders by way of a pressure regulating control panel and a hose assembly line.

Under the "Mother/Daughter" concept, the gas can be received at the main transmission line and at higher pressures, in our case a guaranteed 150 psi. The gas is then compressed to 3600 psi and fed directly into storage tubes mounted on a trailer. The carrier is delivered to the retail outlet and can remain there as the outlet's prime storage or the gas can be delivered into special storage tubes located at the outlet. Either storage arrangement can be connected to the outlet's gas dispensing equipment and the dealer will be charged for the gas through a metering system attached to the carrier tubes.

While the "Mother/Daughter" concept involves high investment, particularly for the truck, trailer and tubes, it is economically viable. It gives us flexibility in marketing, as we are not tied down by the pipeline system, and allows us to complete our objective of establishing a "chain" of CNG outlets. The "Mother" station area could also be utilized as a grass-root CNG retail outlet.

To improve on the economics of the "Mother/Daughter" concept, we are presently investigating the feasibility of a mini-"Mother/Daughter" station which would have the "Mother" station located at a large retail outlet and the trailer and tubes scaled down in size. This would reduce costs by not requiring land purchases/rental and building construction.

The selection of compressors and other equipment will relate to the individual requirements of our program. In view of the high costs involved, detailed studies of pipeline pressures, outlet restrictions and marketing objectives are most important.

Retail Program

To give you an example of a possible retail program I will briefly outline our program. Caltex will supply, on loan to the retailer, all equipment required for CNG dispensing as well as shoulder all installation and related costs. Dealer supply agreements are essentially back-to-back with terms from our direct supplier.

The Caltex price to the dealer for CNG will allow him a margin approximating his present margin for equivalent petrol sales. Caltex revenue is derived from the difference between our contracted costs from natural gas suppliers and our dealer price.

Unlike most competitive CNG outlets, specially designed Caltex pumps will be located on pump islands and the forecourt rather than side or back locations.

The Caltex star on banjo signs at converted stations will temporarily be replaced with a newly developed Caltex Natural Gas logo. We are still studying signing and it is anticipated that after six months, one face of the banjo sign will revert to the Caltex Star sign and that the sign face will rotate. All CNG revamped stations will be painted, if required.

In addition to any dealer solicitation, Caltex will have special gas sales representatives who will call on fleet accounts to sell the conversion program and to direct fleets to a Caltex CNG outlet for its supplies.

Special training programs for dealers and attendants are established to train them in handling, dispensing and promoting CNG and installing conversion kits. Dealers and attendants are also taught simple compression maintenance and are cautioned to avoid tampering with meters, compressors and other dispensing equipment.

Public Relations, Advertising and Sales Promotion

The CNG program is more than just introducing a new product. It is essentially the introduction of a completely new automotive fuel where no market exists. Additionally, it is a fuel about which there is little or no general knowledge. Misconceptions regarding explosion and fire are heightened by the negative environmental aspects of compressor noise and gas storage in outlets.

As a consequence, a well conceived public relations program is a must for both the pre and post-announcement period. Because of the importance of this aspect of our total program, Caltex engaged the temporary services of New Zealand's foremost public relations firm.

For pre-announcement activities, we considered and listed those individuals and organizations we felt would have special interests in the program and who, for reasons of sensitivity, should be informed prior to, or concurrent with, the public announcement. Communications were tailored to individual groups and in some cases personal discussions were arranged by senior Caltex executives.

Examples of those who received pre or concurrent announcement information were:

All Company Employees - Because of the confidentiality of the project in its conception stage, most staff members were unaware of our CNG activities.

Letters to the staff explained the necessity of confidentiality and emphasized job security and job opportunities involved in the project.

All dealers - Letters to dealers were our first opportunity to begin our dealer appeasement program. Only about 50 outlets were to be selected and we were aware that there may be unhappy dealers among those not selected. We pointed out necessary criteria for selection, such as location near gas lines, required pressures, large area required, etc.

Motor Trade Association - This is our national dealer organization. To them we emphasized our election of using present outlets as opposed to opening a separate chain and that with our major program through existing outlets the MTA membership would be strengthened.

Federation of Labour - This organization is the national organization of all individual unions. We informed them of the positive effects on employment that our program would have.

Political Opposition - The two opposition parties were fully informed of our planned activities with special emphasis on the national interest aspects. As the New Zealand CNG conversion policy was set by the present government it was important to us that our retail program not be viewed in a political sense.

In the post-announcement period we sent letters to all Members of Parliament representing areas where CNG outlets were to be converted, the local body authority in those same areas, all national government departments, transportation organizations and all major Caltex consumer accounts.

Most of those with whom we communicated formally acknowledged our communications and I'm pleased to say we received no negative responses.

Conscious of our role in essentially developing a new market, the public announcement of our plans was the most important element of our public relations program. The basis of the announcement was our officially informing the Minister of Energy of Caltex's plan to establish a chain of retail outlets to dispense CNG. This announcement was made at Parliament House and energy and business reporters from the media were invited to a press conference.

From the press conference and media handouts, Caltex's CNG project received national radio and television coverage as well as coverage in all major urban newspapers, trade magazines and most local newspapers.

We have followed up this initial thrust with a series of public relations bulletins on the status of our CNG activities highlighting special equipment to be used, locations of our first 10 outlets, artists' impressions, safety and benefits of CNG.

A well conceived and implemented advertising and sales promotion campaign is vital to the success of the CNG program.

The initial phase of the Advertising and Sales Promotion campaign is designed to create a high awareness of the product, its benefits, the safety aspects and the national interest. The aim is to influence fleet operators and, as a secondary target, the motoring public.

Phase Two will be aimed at generating the maximum number of sale leads from prime potential users.

Following the establishment of a solid base of users (mainly fleet) we will turn our attention towards maximizing sales through our established outlets. Sales representatives' efforts will be primarily aimed at convincing fleet operators to convert their fleets to CNG and to purchase from our outlets. The subject matter is technical and requires demonstration of usage and cost efficiency. A 15-minute sound colour film has been produced to be used as a basis of presentation to fleet operators. A flip chart will be used to highlight features of the film and a high quality color brochure left with the prospect. A further color brochure will be produced to highlight technical features of cost efficiencies and payback period.

Capital Investment and Costs

Even with the Government incentives of cash grants and quick tax write-offs, the development of a chain of retail outlets is costly. In the final analysis, these Government incentives were not required to make the project viable, but they were a significant contributor to the payout and discounted cash flow-rate of return.

Time does not permit a full presentation of the economics of establishing a CNG chain. Costs vary according to equipment selected and outlet volume objectives. The Caltex resellers is expected to service 300 to 500 cars per day and our equipment selection is aimed at this goal. Most outlets presently in operation in New Zealand service 80 to 150 cars per day. Their investment costs are lower but their potential earnings are equally lower, as is their ability to compete with an outlet capable of servicing higher volumes.

Costs escalate with the volumes of cars to be serviced. However, the investment does not increase proportionately. The more cars serviced, the more efficient is your investment cost.

The "Daughter" station costs less than an "On-Line" station, but the additional costs involved in establishing a "Mother" station to feed the "Daughter" makes the "On-Line" outlet the more economic investment.

You should remember, however, the "Mother/Daughter" concept permits delivery in areas where there are no gas pipelines available or where pressure is insufficient to support an "On-Line" station. As your supplies are purchased directly from the main gas transmission lines, your basic product costs will be lower. You should recover full costs of delivery, including depreciation, from your retailer. And a most important advantage is that your competitors can not easily invade your market unless they are prepared to come into your established market at similar or higher investment cost.

In a breakdown of cost elements the largest expense at an "On-Line" outlet is for compressors, while for a "Daughter" station, no compression is involved. Relative to costs for tractor, trailer and storage tubes, compressor costs are low at a "Mother" station.

While costs are high, each "Mother", "Daughter" and "On-Line" station can be fully justified economically. The costs are high, but so are the rewards.

SPECIFICATIONS FOR A RETAIL SYSTEM

Lloyd G. Brown

President and Chief Executive Officer

Welgas Holdings Ltd.
New Zealand

With the increasing discovery of natural gas around the world, current interest is being displayed in compressed natural gas (CNG) as an alternative fuel. This chapter mainly documents the procedures that our company, Welgas Holdings, has adopted and the procedures we have encouraged our future consumers and customers to adopt in planning CNG compression stations. In addition, I will describe results of the dynamometer tests, that we have conducted on the Fiat diesel engine, which are of some interest in Pakistan and Egypt.

Pakistan is rich in natural gas but has no coal and very little hydro electricity capacity and has no capacity for the production of petrol. LPG is desperately short in this country and as a consequence Pakistan has already embarked on a program of development with compressed natural gas as the alternative transport fuel.

Egypt, although well endowed with 70% self sufficiency in crude oil, has large gas fields with more currently being developed. The rationale is that they export the crude oil and utilize the natural gas as Egypt has little ability to generate overseas funds from other sources.

In South America, Thailand, and a number of other South Asian countries, we are finding increasing interest in the adoption of CNG, not only because of the efficiency of the fuel, but also its safety record and the environmental considerations.

As a young soldier I was privileged to serve in Italy during World War II and at that time I saw the initial attempts by the

Italians to use CNG for fuel for transport vehicles. In fact, we captured one of their tanks that used CNG so it shows you the extent that they used it in Italy. With the industrialization of Italy, and the gradual depletion of the gas fields, CNG became less attractive. Also, the quality of the gas deteriorated because of reinjection of water into wells which subsequently shows up in the CNG cylinders on the vehicles. Now the government realizes that Italy must purchase gas - they buy LNG from Algeria, they buy gas from Holland and they also buy a considerable volume of gas from the Soviet Union. This is a deterrent.

The cost of CNG in Italy relative to the cost of petrol should nevertheless be attractive to the consumer. Petrol, at the time of writing, was selling for approximately 930 lira a liter. CNG was selling for roughly 500 lira a liter equivalent but the interesting thing is that diesel was selling for only 430 lira per liter, much below the price of petrol.

Regulations With Respect to CNG

The introduction of compressed natural gas as a viable fuel for transport purposes has led to the review of standards and regulations in order that CNG may gain the widest possible acceptance by the private and business motoring public. In the early marketing assessments, it was realized that CNG refueling stations would have to be located on central city sites for public access and fleet refueling situations. As the most suitable city sites are already retailing transport fuels, it was essential that any new legislation take advantage of existing facilities.

In addition to the mandatory inspection procedures associated with a CNG station (covered by Gas Industry Regulations), it is essential that all related inspections and commercial personnel have a clear understanding of the main functions of CNG refueling so that prospective operators receive constructive advice leading to practical and safe installations. Some of these regulations would not apply when a gas-powered engine is installed in lieu of an electric motor on the compressor.

We find, as I am sure you will find in some parts of Canada, the United States, and other parts of the world, that weird interpretations are made by local body inspectors having a complete lack of knowledge. I recently ran into a typical example of this where one local government officer said that he wasn't going to have a compression station pumping the gas with its noxious smell. I also have adopted the tactic of speaking of pressure, in terms of 200 bars or something to that effect rather than 2,300 psi. It seems to make them quite happy.

Separation Distances

The most important factor influencing any potential site for CNG refueling is the space available for locating the primary items of equipment.

All gas cylinders up to 250 liters water capacity are governed by the Dangerous Goods Regulations administered by the Department of Labour. While the wording is somewhat different between the New Zealand Standards 5425 and the Dangerous Goods Regulations, the requirements are the same. For the purpose of this exercise only gas storage units up to 4500 liters total are explained. We don't go beyond that in our gas storage.

The minimum horizontal distance between the gas storage unit and any building not protected by a four-hour fire rating wall is 2.5 meters. This is considerably less than one would ever be permitted to have in Italy. When a four-hour fire wall is available, the separation distance can be reduced to one meter. Vertical distances are governed by the delineation dimensions shown in New Zealand Standards 5425, Part I, pages 19 to 21. Appended to this chapter are the two standards that have already been published and accepted in my country - New Zealand Standards 5425 and 5422 Codes of Practice, dealing with CNG compression and refueling stations; secondly with the use of LPG and CNG fuels in internal combustion engines. I recommend that those of you who are involved in the various standard committees make yourselves familiar with these to avoid going through the anguish that New Zealand went through. Normally, you publish standards relative to the gas industry in the international color of yellow. These are printed in red to represent the blood, sweat and tears that went into them.

The minimum horizontal distance between gas storage units and any pedestrian walkway or other public place is three meters unless separated by a four-hour fire wall. In these situations, gas storage units containing cylinders larger than 250 liters capacity shall be separated by five meters from a public place.

The minimum horizontal distance between a gas storage unit and a liquid fuel dispenser, which is petrol, kerosene, etc., is five meters. Therefore, you can put your gas filling probe on the station with little restraint.

There are no specific conditions relating to separation distances in relation to buildings for gas compressors except for the requirements of local bodies, and it is also necessary to conform to the electrical regulations in New Zealand. I know that this applies in various forms throughout Canada. For example, Alberta probably has the most stringent electricity regulations seen anywhere in the world.

As CNG is a new development, potential operators should be encouraged to seek early advice from the local gas utility in respect to siting a compressor and its relationship to other components that require additional inspection procedures, and that relates to electrical and dangerous goods. I, as one in the industry, accept the responsibility for advising potential compression station operators of their responsibilities in all of these matters. It is not enough to advise the prospective enquirer to go to the local electricity office, or something like that. We should give him all the advice we can and then send him off to those authorities.

The standard regulations and codes relating specifically to CNG do not contain dimensions in relation to noise level separation. Most local body regulations do place limits on maximum noise levels except at site boundaries and the Department of Labour also administers regulations governing noise control in other work places.

With respect to the provision of fire walls and security fencing, the main requirement is to permit reduced separation distances between gas storage units and buildings, public places, etc. Such walls must be designed for a four-hour resistance rating and will usually be constructed in reinforced concrete or reinforced concrete blockwork.

Gas storage units require security fencing to all elevations not protected by any wall or building element equal to the height of the gas storage unit.

Security fencing must be a minimum distance of one meter from the gas storage unit, and gate access must be lockable. Further protection for gas storage units located adjacent to vehicle areas must be provided to resist impact from vehicles. This additional protection usually is in the form of an elevated concrete slab base forming a kerb at least two meters from the gas storage unit.

In assessing a site suitability for CNG refueling facilities, it is necessary to consider the relevant approval procedures for licenses and permits associated in such an operation.

The following is a list of procedures which will generally apply to most situations. Throughout Australia, New Zealand, Canada, the United States, and other countries where we have become involved, I don't find too much difference in gas regulations nor in local body regulations excepting where there are extreme conditions as in my own city of Wellington - or perhaps Los Angeles where there are quite severe restraints with regard to seismic activity.

i) Check with local gas utility to ensure availability of gas volume and supply pressure.

ii) Check with local electric power supply authority regarding starting conditions for electric motors and advise the authority of horsepower requirements.

iii) In some districts it may be necessary to obtain "approval in principle" from the Territorial Authority before proceeding with detailed plans for the installation. If building alterations or extensions are necessary, a "building permit" will be required.

We find this particularily in some of the rural areas. As you are probably aware, we only have gas in the North Island of New Zealand. The pipeline runs from Plymouth, both north and south, to the cities of Auckland and Wellington. We are at the moment proposing several branch lines. There is one going to Hawke Bay and one to Tauroa. Generally speaking, there will be a number of compression stations in rural areas.

Now, these areas do not have formal building regulations and they either rely on the Dangerous Goods Inspector or some other local authority to advise them in these matters.

iv) Formal application for a gas supply must be made with the gas utility by a person registered under the Plumbers and Gas Fitters Registration Act.

In New Zealand, this is a peculiar thing which, I am hopeful, may be changed. I think it is an unnecessary restraint. I believe that there are many people in the industry supplying compressors with just as much experience and ability and competence as there is in the gas industry. I think that to put such an imposition on them is quite unfair. However, I think it does protect the gas utility. You do know who is installing a compressor.

v) "Dangerous Goods Licence" is required for the gas storage units. This licence would normally be considered during the building permit approval stage and form part of the total fee paid to the Territorial Authority. However, it is advisable to apply for the Dangerous Goods Licence before proceeding with the installation.

vi) If the installation is eligible for a Government Gas Development Grant, it is necessary to make a preliminary application with estimated costs etc., before ordering any equipment.

I sincerely hope that before too long there will be a requirement in your regulations indicating that your governments will back CNG.

Further aspects to be considered are conditions adjacent to properties. Locating CNG stations on central city sites will generally require designing and locating the primary components with the maximum allowable separation distances. In these situations, care must be taken to ensure that doors, windows, and other openings in adjacent properties are outside the delineation division or separation distance boundary. This also applies to discharge ports on gas vent pipes, which can easily be overlooked during the final commissioning stages of a compressor installation.

On some of our very congested, already existing petrol or gasoline retail outlets, it is quite a problem to get the vent pipe away from the building.

Pressure regulating stations (PRS) need more attention, and this is often overlooked in the initial planning stages. Experience to date has shown that most PRS units for CNG stations are somewhat larger than other comparable commercial installations due to the additional requirements for a pulsation damping chamber.

Vehicle Conversion Equipment

Now I would like to turn to the vehicle conversion equipment. Earlier it was commented that we were the doctors to the motor industry and we had to be proficient. In my opinion, we can relate the three major components of CNG equipment on the motor vehicle to the human body. In this respect the cylinders are the lungs, the kit is the heart and now, with the development of electronic spark control, we can consider that the brain. Fitting the largest sized cylinders and tailoring the cylinders to the dimensions of the trunk of the vehicle with minimum intrusion into the space, the vehicle can travel satisfactory distances after each refueling. For this reason, we have concentrated on what is known as the "Normale" cylinder produced by Faber Company of Northern Italy. These cylinders have the greatest capacity for the least weight: the 60-liter cylinder weighs 54 kilos. This company intends to produce cylinders up to 80 liters capacity that will, as I have stated, weigh less than one kilogram per liter of fuel.

I can speak as one who operates a fleet of between 50 and 60 vehicles, consisting of cars, vans and trucks. We find that when we use these cylinders on the Ford transit type vehicle with two 40-liter capacity cylinders, our gas industry service vehicles can travel for at least two days without refueling.

Now, I am fully conscious that what we can do in Wellington is probably different from what can be done in Auckland. I am equally aware that in cities such as Calgary and others, the distances that service vehicles travel vary inversely with the size of the city. However, in travelling around the world I have found that most taxicabs, whether they are in London, Washington or Wellington, travel about 125 to 150 miles per day. Indeed, in London they would regard 100 miles a day as quite abnormal. Now, that indicates that we can really penetrate the taxicab industry because those vehicles are obviously working under unrealistic conditions for what is the power of the vehicle of today.

We all know that a petrol engine does not operate well when it is in stop-go conditions and yet that is where the performance comes out with the CNG-powered vehicle whether it be compared to petrol, gasoline or diesel.

Again, as an operator, I insist that the fueling location in a converted car be in a preferred position as close as possible to either the front left hand or right hand of the vehicle immediately under the hood. This means that there is no excuse for the driver not to check the condition of the fan belt, the level of water in the radiator, or the level of oil in the engine sump as well as the condition of the battery. After two and a half years continuous experience, my company is finding that this is having a definite effect on our maintenance costs.

Now I'd like to turn to kits. The first kits employed in New Zealand in any numbers were the well-tried and proven Dual Fuel System kits of Southern California. These kits had established an excellent safety record as well as performance on the road. They had one disadvantage and that was the cost of the equipment. The Dual Fuel System is equally acceptable on a range of vehicles from small cars to the larger commercial transport vehicles. However, we did find, and I am sure my friends from Dual Fuel would agree with me, that it becomes a little bit more expensive to fit them to the new profile of the small compact cars. We have now concentrated on Italian equipment and we have preferred the Renzo Landi equipment. This equipment is now produced in two distinct sizes and so far has been found to be adequately acceptable on all types of vehicles. We have to make some modifications on the diesel engine to give it the greater capacity.

The new Dual Curve electric control gear system currently

being produced by the Autotronic Corporation of El Paso, Texas has eliminated the disadvantage that was so loudly trumpeted by the critics of CNG with regard to power loss in petrol engines. Again with some considerable experience, I have always considered that a power loss in excess of 7 to 10% indicated incorrect tuning or installation of equipment. Since we have had access to the new Dual Curve ignition equipment, we have been able to eliminate all of the losses and in fact, improve the distance travelled when on CNG.

Now, I am fully aware that there will be many other makes and suppliers of this type of equipment to deal with one of the problems that we are all facing.

CNG and Diesel

With respect to CNG and diesel engines I have been working on the application for several years. This technology has been known for some time but has not been available in a satisfactory form for the demanding conditions of road transport. We in New Zealand, in association with an Italian company, have now overcome these problems and have developed kits suitable for application in the various configurations in diesel engines. Our equipment allows us to tune the engine to accept a 50% natural gas and 50% diesel mixture - or as little as 20% diesel and 80% natural gas.

As Professor Karim has indicated in this volume, you can, under certain conditions, maintain compression ignition with pure methane but you do certainly have to do something about it. I think from both operational and commercial points of view we should start out with a target of 20% diesel and 80% gas because of the power advantages.

We have done a number of vehicles with quite different engine configurations and I am encouraged that the data that have been produced are comparable to our tests in Italy and New Zealand. This refers particularly to the exhaust temperatures and the operating temperatures of the engine.

Technical Training

One of the most important, though not very exciting, subjects with regard to the development of CNG, and one that is frequently neglected, is the training of technical staff. The adoption of CNG is a totally new technology in the automotive industry and places a strain on the personnel involved. For a number of years in both the United States and New Zealand I have been directly involved in the training of technical staff. One of the great challenges facing the motor trade and the gas industry is to have a viable, well-trained staff to meet the needs of this new developing industry.

With the help of both the Minister of Education and the Minister of Energy in New Zealand, we have been able to develop training programs for the auto mechanic already employed in the industry with regard to the conversion of motor vehicles. We are fortunate in New Zealand that under our technical training regulations and our apprenticeship schemes, motor mechanics who have a formal apprenticeship must sit examinations in order to get certification as "A" Grade mechanics. This gives them the right to provide vehicles with warrants of fitness and they can go on from a normal "A" Grade mechanic to become either an auto electrician by studying under another module; or a diesel mechanic, or a specialist in automotive gear boxes, etc. So we have the infrastructure to train these people. What we have done is simply impose another module on top of the existing training program so auto mechanics can become trained in the installation of both LPG and CNG on motor vehicles.

Another area that has been neglected, in my opinion, is with regard to the ancillary equipment. I am talking about not only the meters and regulators, but with particular attention to the compressors. In my opinion, more attention must be given to this area as the staff employed in the gas industry by gas utilities cannot be expected to have gained the necessary experience in the maintenance of compressors. When one considers the wide range of compressors that are being offered, it is obvious we must provide training courses in the use of all of this equipment.

Now, in New Zealand we have compressors from Sweden, from England, from France, from America, from Italy, and from pretty well every country that makes compressors around the world. We have also got the same problem with regard to kits. In fact, I've discovered that there is an Italian kit on sale in New Zealand that is not on sale in Italy! I think this is healthy but at this stage of development of the industry we must be very careful that substandard equipment is not allowed to be installed. You can imagine the ball that our critics would have if they could find anything like that to hang on to this industry.

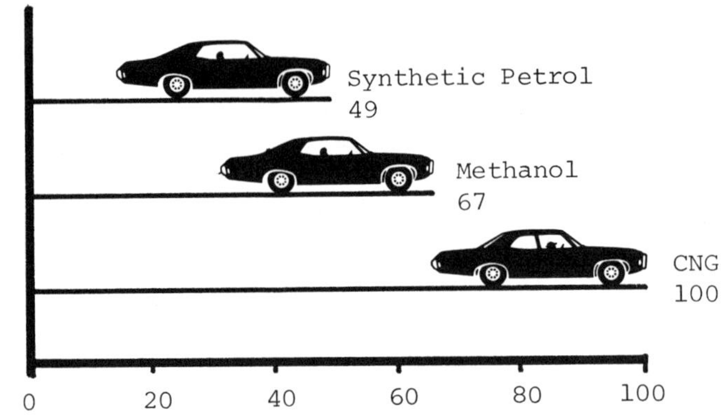

FIGURE 1. Distances Travelled on a Unit Amount of Maui Gas used to form each Fuel

FIGURE 2. A New Zealand gasoline pump

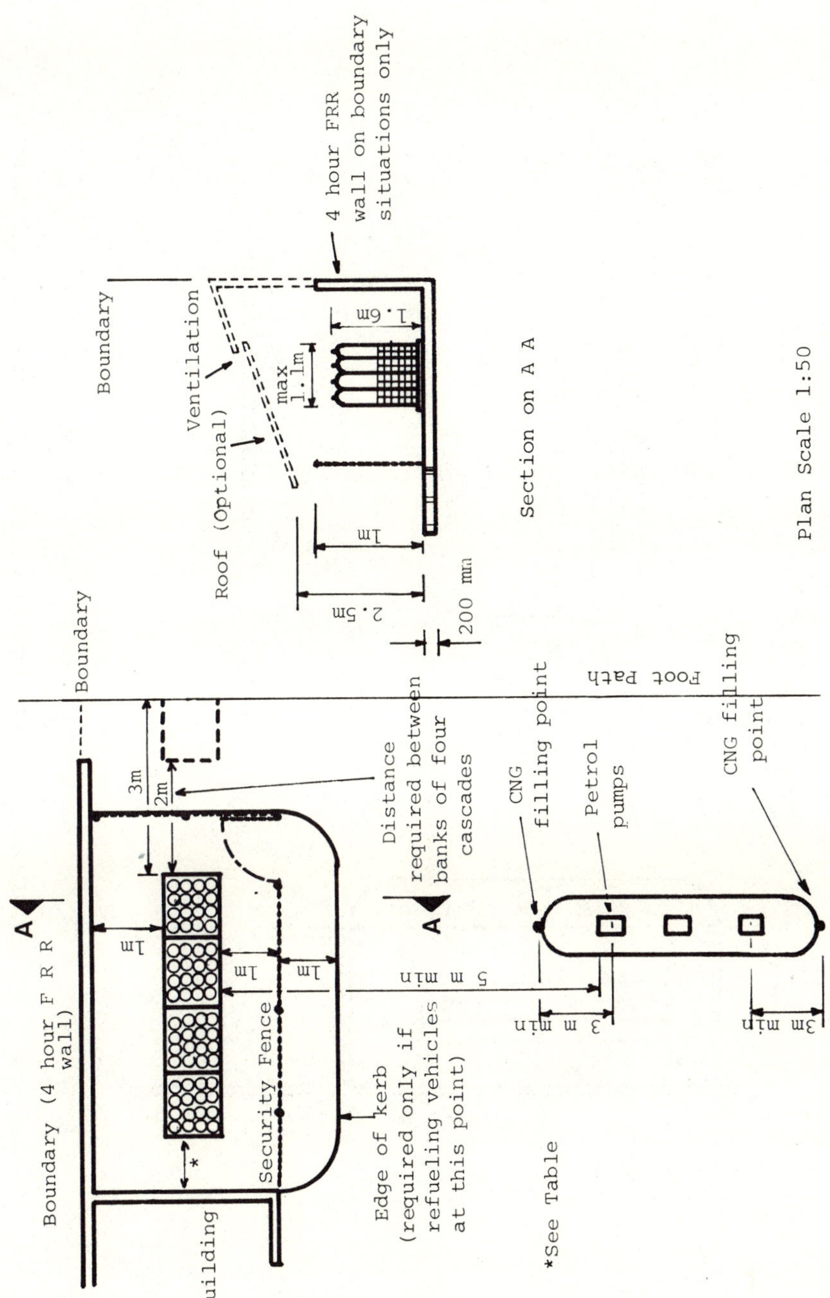

APPENDIX A. Typical Layout of CNG Storage Units: Separation Distances (a)

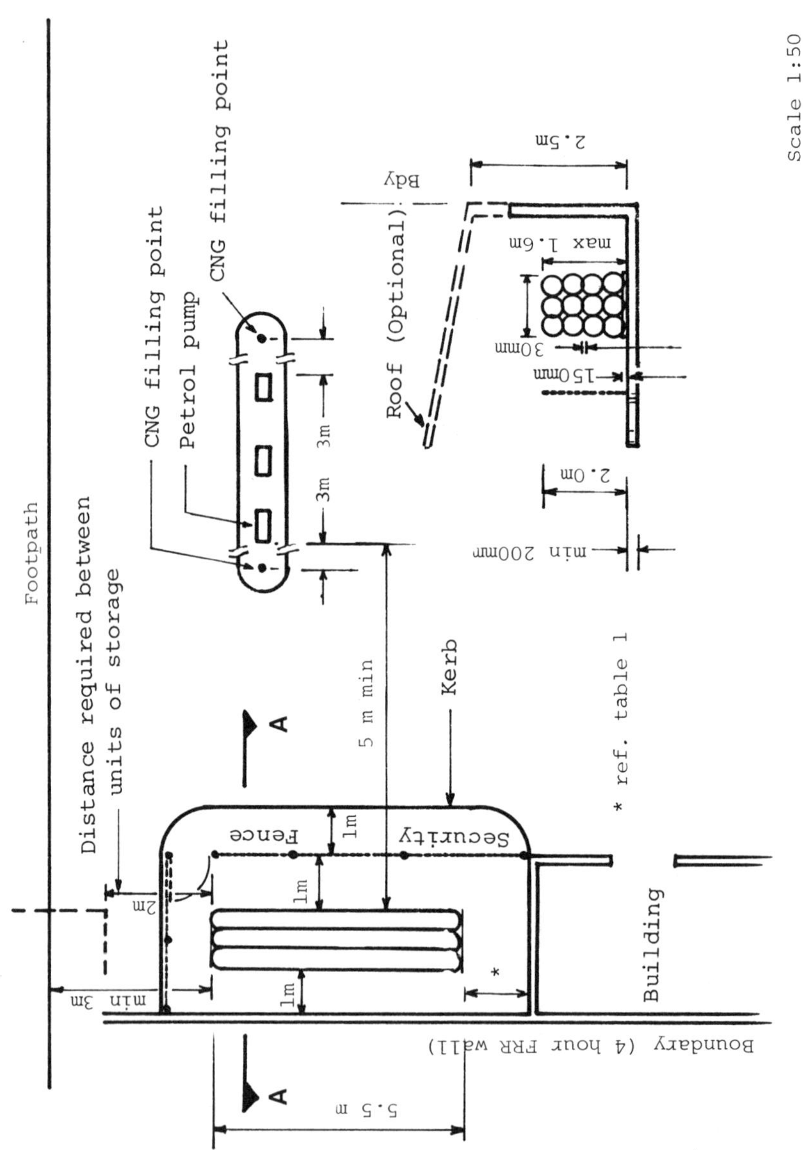

APPENDIX B. Typical Layout of CNG Storage Units: Separation Distances (b)

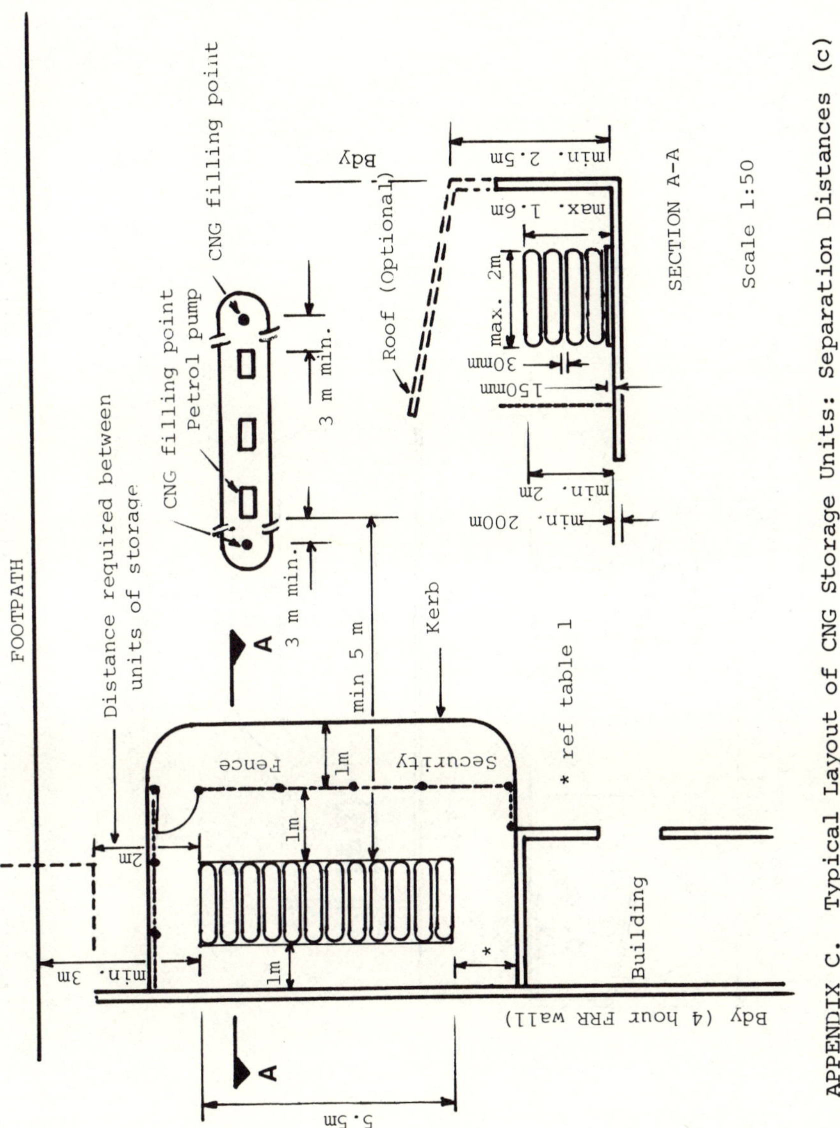

APPENDIX C. Typical Layout of CNG Storage Units: Separation Distances (c)

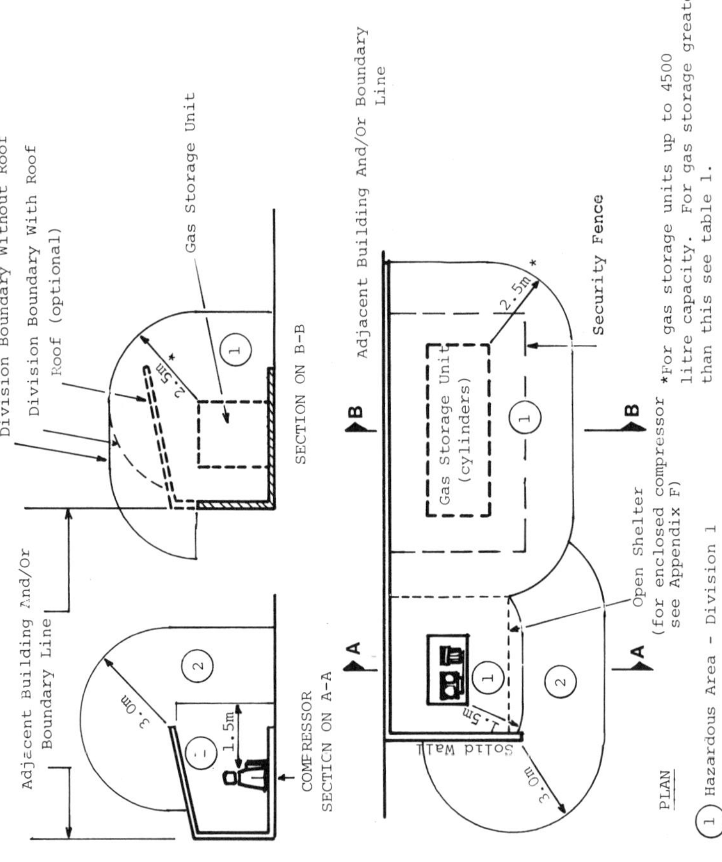

APPENDIX D. Hazardous Area Delineation (Adjacent Building)

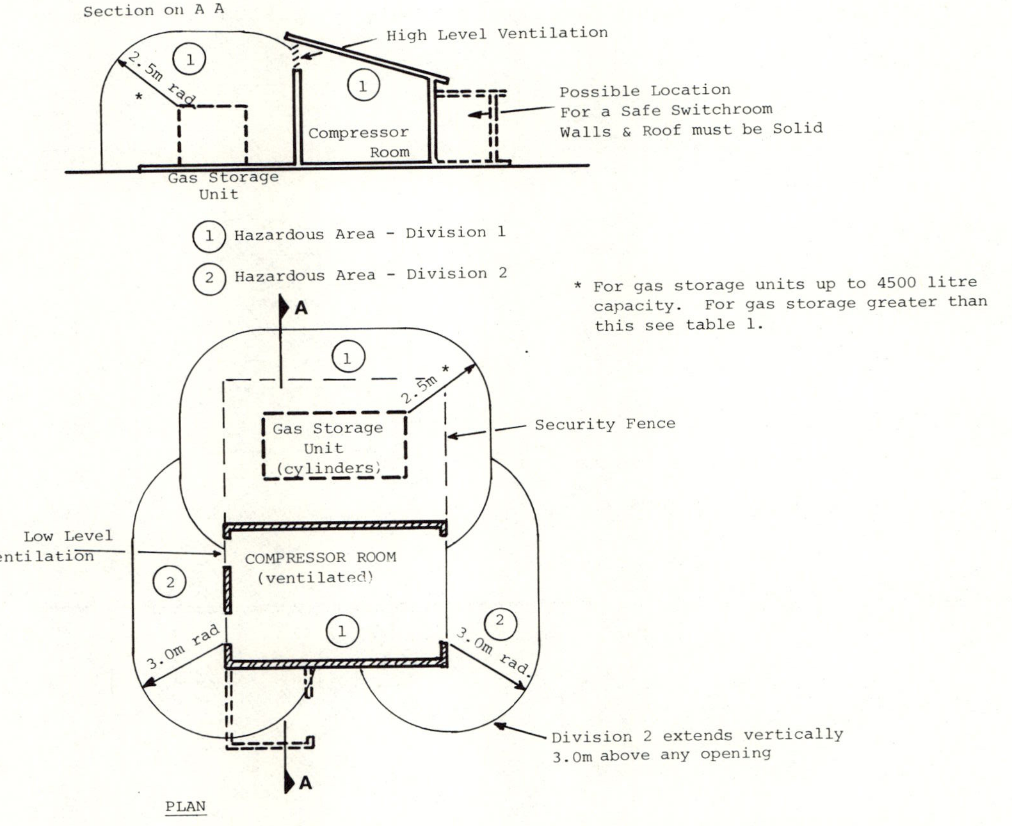

APPENDIX E. Hazardous Area Delineation (Enclosed Compressor)

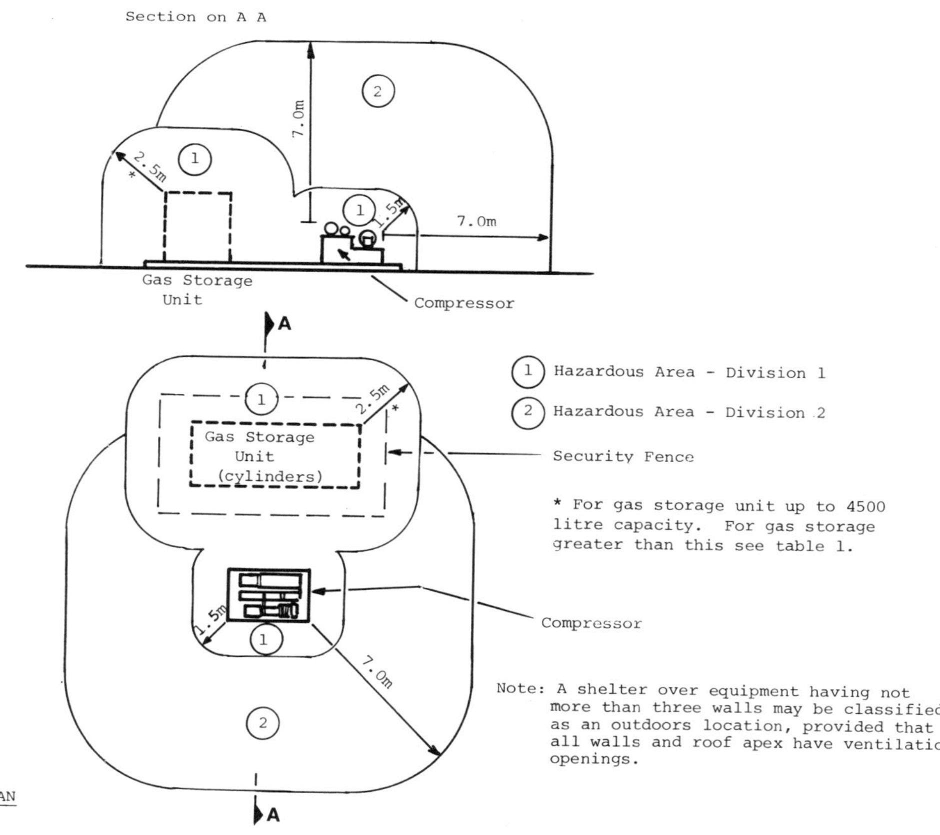

APPENDIX F. Hazardous Area Delineation (Outdoor Installation)

THE EXPERIENCE OF CNG FUEL SYSTEMS LTD.

J. Judd Buchanan

President, CNG Fuel Systems Ltd.

Calgary, Alberta

Ladies and gentlemen, I stand before you not as an engineer, nor as a research scientist, but rather as a businessman who also happens to have some political background. I am the president of a company, the first such in Canada, that has decided it now makes hard business sense to move aggressively into large-scale conversion, across this nation, from liquid fuels to compressed natural gas for motor vehicle use.

I am therefore pleased with the double focus announced for this conference by the Honourable Patrick McGeer during his speech last June 15th to the Institute of Gas Technology Symposium held in Detroit, Michigan. At that time, he announced that we would be gathering here in Vancouver "to assess the current state of the methane option and discuss the basis for a global strategy". Many of you are well placed to assess the current state of the methane option. We at CNG Fuel Systems Ltd., however, have been hard at work on our own implementation strategy for that option, and it is that which I should like to discuss with you now, along with the background analysis which led us to the conclusion that the time was right for such action.

Having just promised not to attempt to teach you about the properties of methane, I shall, at least, summarize their principal advantages; the advantages that make CNG such an attractive fuel option, especially as compared to gasoline. It is in abundant supply, inexpensive to process and easy to transport. It is clean, not only for the interior of the engine but, equally importantly, for the environment. It is non-toxic and safe. It is less expensive than gasoline, giving a substantial and growing economic benefit to users.

Even in this brief resume, supply deserves more emphasis than I have so far given. In the words of Richard Wyman, our Vice President, who is also Director of research for Canadian Hunter Explorations Ltd., "world-wide supply appears overwhelmingly promising". This is especially true in Canada. We are sitting on top of an enormous reservoir of this efficient, clean, safe, inexpensive and available fuel.

However, as we all know, the corner "gas" station is in fact selling liquid gasoline, not gas. Despite all the advantages I have just mentioned for using natural gas rather than gasoline, we have been, and still are, dependent on oil, not natural gas; oil, I might add of which Canada must import 30% of our annual supply, as opposed to natural gas, of which we have an abundant supply.

The "why" of this situation goes back to business decisions made at the time that automobiles were first being manufactured. The internal combustion engine of the 19th century did, in fact, run on gas, not gasoline but, when it came time to put the internal combustion engine into an automobile, there were neither the compression techniques nor the storage techniques capable of providing the compact fuel form that was required. And so the auto industry and its technology evolved around liquid fuels - first gasoline, later diesel.

This situation reinforced itself over the decades; the industry acquired its habits and so did the customer. CNG technology also evolved - Italy led the way in the 1920's with new compression equipment and the introduction of CNG-powered vehicles and Italy has had them ever since - but despite this, we remained massively wedded to liquid fuels. There was nothing pushing us to break out of the closed circle - the CNG fueling stations didn't exist because cars designed to use CNG didn't exist; the cars didn't exist because customers didn't call for them; customers didn't call for them because there was no place to get the fuel. And anyway, gasoline was cheap.

It is true that in the 1960's we became more concerned about the environment, but the ecological advantages of methane did not provide sufficient push for mass conversion to take place. Take the example of Dual Fuel Systems, a subsidiary launched in Los Angeles in the late 1960's by Pacific Light. It was an act of good corporate citizenship -- Pacific Light's clean fuel contribution to the fight against Los Angeles smog. As the name suggests, Dual Fuel Systems was promoting the conversion of cars to run on dual systems, i.e. liquid fuel and natural gas. The company, however, made little headway until the last few years following 1973 when a new four letter word was added to our vocabulary - OPEC. Oil prices began to balloon and, even though natural gas prices have to some extent tracked that rise, the absolute dollar gap between gasoline and natural gas has been doing nothing but widening ever since.

The crucial transformation factor is economic, always economic. People may worry about the environment, may - especially after experiencing gasoline shortages as they did a few years back in the United States - worry about security of supply. But the cold, hard truth is that the motorist has to be convinced that it is to his or her economic benefit to switch before that switch will take place. In other words, we're talking about the pay-back period for the conversion. In our company, we're operating on the premise that people have to see a two year or less pay-back period before they're going to be interested. That day is fast approaching, for even the average motorist, and a few other recent events have brought it even closer.

Those other events, and I am here speaking in specifically Canadian terms, are last year's National Energy Program and the recently concluded energy agreement between the federal government and the province of Alberta.

CNG Fuel Systems Ltd. was incorporated on August 15, 1980, as a private company with four shareholders; three men from Canadian Hunter Exploration Ltd., John Masters, Jim Gray, Richard Wyman and myself. It was their baby - they had started moving on this even before the NEP and subsequent events gave new impetus to the concept. The three men first became interested in the idea because they were interested in finding markets for all the closed-in gas which the industry and, of course, their own firm, had in Western Canada. Our supplies are very great. Consider the following situation, and how frustrating it is from a businessman's point of view.

Canada is presently importing approximately 500,000 barrels per day of crude oil, while surplus natural gas is shut in. The annual cost of importing that oil in 1980 figures-and it will have risen sharply by 1985-was a staggering $8 billion. Meanwhile, for the last five years Canada has been adding over 5 trillion cubic feet per year to her proven reserves of natural gas, yet in 1980, only 1.6 trillion cubic feet were being used domestically, and less than 1 trillion cubic feet exported. We have every reason to assume, given markets and reasonable returns to the producer, that the industry can continue to add at least 5 trillion cubic feet per year. In fact, by 1990, a more realistic figure, given contributions from the Frontier areas, would be 8 or 9 trillion cubic feet. Yet by that same date, 1990, the National Energy Board estimates that less than 3 trillion cubic feet will be needed within Canada, even with expanded domestic markets resulting from the proposed pipelines into Atlantic Canada. The three men from Canadian Hunter, therefore, were keenly interested in developing new markets through the introduction of new uses for natural gas. And so ...CNG Fuel Systems Ltd. was born on August 15, 1980 - the same day I joined the group. My colleagues intended to make money, but they were also convinced that a conversion program was vital to Canada's economic and security interests - that it could and would play a vital role in moving us to energy self-sufficiency.

Events since August 1980 have only served to make CNG a more attractive option than ever as a motor vehicle fuel. Now before my good friends in the Calgary oil patch, who have strong reservations about the recent energy agreement, slit my throat, I must carefully say that I am going to speak only of what's good for CNG, as a company, and for CNG as a motor vehicle fuel.

The National Energy Program, introduced in October 1981, did not make the oil industry happy, but it did give official federal recognition to the importance of CNG to the future of Canada. The government stated that it was interested in the potential of CNG as a motor fuel and that it would encourage the development of this option. It outlined certain incentives, such as support of demonstration fleets and the necessary fuel facilities through existing federal-provincial Energy Demonstration Agreements and, as necessary, directly by the government of Canada. It is anticipated that incentives will soon be provided for CNG along the lines of the propane conversion incentives, which provide for taxable grants of up to $400 for each vehicle of a commercial fleet converted to propane.

That was - again I stress, from a strictly CNG point of view - encouraging. But what really paints a very attractive future for CNG is the recent energy agreement between federal and provincial authorities. We're now looking at a $3.50 to $4.00 gallon of gasoline within 4 to 5 years. By then, CNG will cost between $1.50 and $2.00 per gallon at public fueling facilities and even less for fleet operators. The average motorist, who puts 12 to 13 gallons into the tank per week, will see that $2.50 per gallon price spread every time he drives into the service station. Two and a half times 12 gallons .. that's $30 a week, which starts to be a compelling amount of money. At that price we're starting to look at approximately a one year pay-back on conversion for the average motorist. So we feel confident in stating that the individual motorist is going to be interested in compressed natural gas, and in the very near future. However, as Pat McGeer rightly pointed out in his Detroit speech, we need a "global strategy". As of today, that potential CNG customer can't act on his interest - he can't convert his car, and he can't fill up with CNG at any public service station in the country.

Before getting to the global - or even Canadian - level of discussion, I will outline for you the strategy of our own company.

CNG Fuel Systems Ltd. is a facilitator. We exist, first, to sell the Canadian public on the concept of switching to CNG for their motor vehicles. And, I might add, in one year of operation, we have seen vastly heightened awareness amongst Canadians about CNG as a viable and "here-now" energy option.

We also exist to sell, at the wholesale level, a number of components. Starting in early 1982, we shall be manufacturing hydraulic compressors in our own facility in Brampton, Ontario. We shall therefore be manufacturing, selling and servicing CNG compressors - that vital link in the CNG fueling chain.

As well, we are presently working on arrangements to manufacture CNG cylinders, to be used both in vehicles and in compressor facilities. We don't expect to get into extensive manufacturing of the smaller componentry - but we shall be assembling those items into conversion kits, combining those kits with our cylinders and selling them - again, at the wholesale level - to various fleet owners, to service station owners and operators, to car dealership... in other words, to the people who will be doing the installing in the roughly 50,000 service bays that exist in Canada.

We see another component essential to this whole process, namely that we have to have a certain number of conversion training centers as well, where we'll be doing conversions ourselves. There are two reasons for this. One is that we can then get direct, hands-on experience with all the combinations of vehicles, motors and options and solve any problems which arise. That means that when the individual service station operator, who has been trained to do conversions, calls up to say he has a conversion customer with a 1981 Chev Malibu Classic with a 262 cubic inch engine and two barrel carburetor on his hands, we'll send him the exact appropriate conversion kit for that specific car.

The other purpose for the conversion centers is to act as training centers. We will be offering courses to the people who will be doing the conversion work - the fleet operators, service station operators and car dealership staffs.

We expect that ultimately the courses will be offered as part of the general mechanics course in technical institutes like SAIT, NAIT, the B.C. Institute of Technology, in community colleges and vocational shools throughout Canada. We've already opened discussions along this line with various educational institutions.

We're working closely with our New Zealand friends on arrangements for manufacturing diesel/CNG conversion units. Many of you will have seen the diesel engine in the demonstration area operating on a diesel/CNG blend.

We have also been sponsoring specific research and development efforts at the Western Washington University in Bellingham and the University of Calgary. These initiatives are to help improve the performance of CNG fueled engines. We're convinced that 130 octane natural gas can and must be made to outperform 85 octane gasoline.

Finally, to round off this list of components and services, we intend to offer leasing arrangements, whereby customers can avoid paying large amounts of cash up-front and can instead lease the package and pay for the conversion over time, out of their fuel savings.

You'll notice, here we differ from the Italian model and follow the New Zealand system in that we have no plans for a network of free standing CNG fuel stations of our own.

We looked at it very carefully, and concluded that it just didn't make economic or marketing sense. It would have required massive amounts of capital outlay on our part, acquiring sites, building stations and paying staff; it would have required customers to break ingrained fueling habits and search us out; and it would probably have incurred the opposition of the existing networks of service station owners and operators, rather than gain their support, since they could only foresee losing their customers. We didn't see how we could economically compete with those existing stations when they would have to invest only $150,000 to $200,000 to add a CNG dispensing facility, without any initial incremental labor cost. On the other hand, we would have to spend in the neighbourhood of $1 million to set up each of our own stations, depending on real estate costs, and then staff it exclusively for CNG sales. It didn't make economic sense, and to us it didn't make marketing sense either.

Here is our game plan: we provide compressors, cylinders and conversion kits; we establish a number of conversion centers with a training component to them; and we tie in with existing service station networks in dual system arrangements, whereby they go on offering liquid fuels just as they do now, but add a CNG facility.

New Zealand experience indicates tht a service station which adds a CNG fuel facility also enjoys a substantial increase in their own sales - gasoline, oil, tires, batteries, accessories - motorists like convenient one stop shopping.

There are a couple of very recent developments worth mentioning: two weeks ago, we not only opened our first conversion centre, which is in northeastern Calgary, but also announced that we had reached agreement with Husky Oil, whereby Husky will be setting up three public fueling facilities, the first near that conversion center. Our second conversion center will be in

Vancouver at 1290 S.E. Marine Drive, and it opens for business in the next couple of months. There will be another Husky CNG outlet nearby. In addition, Husky will open a third station in Edmonton, at a site yet to be determined. These are the first public fueling facilities to be announced in Canada, and they will all be in operation by late 1981 or early 1982.

Those are the basic elements of our present activities. We are going about them slowly and carefully ... for we've involved in a whole gamut of chicken-and-egg situations. We intend to develop our own capacity to make certain it works well and properly. We are determined that the car owner must have a positive and happy experience with CNG.

And then, we have a whole other level of activity - and I think this is when I begin to discuss the larger "game plan" that Pat McGeer is urging upon us at this conference - that of government regulations.

Federal, provincial and municipal - each level has its own responsibilities within the broad regulatory regime. We have worked with the federal government in encouraging the Canadian Gas Association to set up two committees; one to look at standards for the componentry to be used in conversions, and the other to look at the installation code. The first committee, after some six or seven months of work, produced a set of preliminary standards. The other committee has also issued a draft supplement of the installation code, dealing with installation in vehicles and in fueling stations. We've been making good headway, but there is still work to be done.

In fact, as I suggested earlier, we see ourselves as facilitators, and I think this is our role in the larger game plan. We have chosen a limited role of our own and we must have many partners in the overall activity. It means that we must work with government at all levels, with natural gas companies, with distributors, with service station operators, and with various educational institutions and research facilities as well. There is a lot of talk going on, and the need for a lot of coordinating of action - but you reach the point when someone must pick up the ball and go for it - that is our role.

It is going to happen. Our immediate target is to see 30 to 50,000 conversions done in Canada in 1982; in 1983, 100,000 conversions; by the middle of the decade, 500,000 vehicles. That is out of a total in Canada of some 12.5 million vehicles - only 4%.

We believe that once there are two or three hundred thousand vehicles on the road running on CNG, the auto makers will feel that a meaningful market exists and it is time to start serving it with the CNG option as original equipment. It's interesting to note, by the way, that in New Zealand, which is a few years ahead of us in its CNG endeavors, some of the best sales organizations for conversion to CNG are new car dealerships which have begun offering conversion kits to the consumer as another vehicle option at the time of original sale.

There are other developments we see, even farther down the road. We want to consolidate in Canada first - that's number one - but certainly we'll then be looking for export markets - either straight export, or in joint ventures. There is tremendous potential. Third world countries have been rocked back on their heels by oil prices, yet in many cases they are sitting on tremendous natural gas reserves of their own. I'm thinking of countries like Malaysia, Pakistan, Thailand, Colombia and Egypt, for example. Obviously, if they could develop and exploit this indigenous fuel source, they would be much better off.

Then there is the whole industrialized world. CNG Fuel Systems Ltd. intends to concentrate in the Western hemisphere, and there is tremendous potential in the American market and in Mexico as well.

To conclude, we see a bright future for CNG as a motor vehicle fuel...not as a fuel that ought to be considered, but one which is being considered; not as something that people should do, but as something they will do. They are going to do it because it will make overwhelmingly good economic sense for them to convert, and because our company will be working with people like Husky and the natural gas companies to bring that option to them. We also see a bright future because conversion to CNG is so clearly in the national self-interest that the prospect of increased governmental cooperation is very great. Natural gas is a secure supply, an abundant supply, increasingly inexpensive in comparison to the liquid fuel option and increasingly available, as gaps in the pipeline distribution system are being filled in.

Conversion can also have enormous consequences for our foreign exchange situation. I said earlier that we have a target of 500,000 vehicles converted by the middle eighties - optimistically 600,000. Those first convertees will be primarily high mileage, high consumption fleet vehicles - taxis, police vehicles, postal vehicles, delivery vehicles. We calculate that their gasoline consumption will be in the order of 1 billion gallons per annum - you have to refine 100 million barrels of oil to yield 1 billion gallons of gasoline. That figure represents two things - firstly it equals Canada's approximate net oil imports and secondly it equals the projected production of both the Alsands and Cold Lake projects with their $25 billion investment requirements.

CNG Fuel Systems Ltd. believes that conversion will be good for the individual Canadian and good for the nation as a whole. The scale of opportunity is immense. Very hard-headed businessmen are backing that premise with millions of dollars in long-term investment. They know they are in for the long haul, but they do not expect to lose their investment.

USE AND DISTRIBUTION OF NATURAL GAS

Velio Bellini
Design Engineer
Reciprocal Compressors for Refuelling Stations

Nuovo Pignone, Florence, Italy

The vast potential of natural gas as an automotive fuel is borne out by the over 50 years' positive experience accumulated in Europe, and mostly in Italy. Loosening and updating of current Italian standards - deemed far too restrictive - are felt to be in the offing. Forthcoming developments in metering equipment will allow even greater simplification of distribution systems, thus favoring their more widespread utilization.

Introduction

One consequence of the worldwide energy crisis has been that many nations have stepped up research aimed at developing vehicles powered by non-petroleum fuels. As petroleum reserves diminish, and gasoline becomes scarcer and costlier, increasing interest is being focused on other fuels such as coal-derived gasoline, alcohol, hydrogen and natural gas.

Of the alternate fuels, natural gas appears to be the ideal solution for the immediate future, especially in countries where there are large-size gas deposits, yet not equally large crude oil reserves. In such cases, use of natural gas not only pays off in economic terms, but also provides the additional benefit of freeing the user nation from the dependence on foreign suppliers.

Compressed natural gas as an automotive fuel is not new to Europeans. Use of CNG started and peaked during World War II all over Europe. Thereafter, its popularity dipped sharply in the cheap gasoline postwar era, except in Italy where it has always been an alternative to petroleum fuels.

Advantages in using CNG

An important advantage of natural gas is that it does not pollute and it is thus an aid in ridding the world's congested urban centers of the petroleum-created pollution plaguing them today. In fact gas, unlike liquid fuels, produces virtually no soot during combustion and, unlike gasoline, produces no lead compounds. Also, other harmful substances such as CO and NO_2 are reduced to the barest minimum.

For this reason, natural gas is an especially appropriate choice for public transport vehicles such as taxis and buses, as they operate primarily inside, or in close proximity to, metropolitan areas, and are thus among the worst urban polluters.

Natural gas possesses all the requisites of an excellent engine fuel. It has a thermal potential slightly below gasoline's and a high octane rating (about 125) that means it can be used in existing engines, even increasing efficiency.

One important benefit from using natural gas fuel is that engine life is much longer than if gasoline is used. This is a result of perfect combustion and ensuing lack of carbon deposits. Oil changes are required much less frequently than in conventional gasoline-fueled engines and, as non-detergent oil can be used, this represents an additional appreciable moneysaver. Moreover, from a practical standpoint, the standard vehicle transformation makes it easy to go from gasoline to natural gas and vice versa whenever desired. Changeover may be effected even when the vehicle is in motion, simply by pressing a button on the dashboard.

Distribution network

Natural gas is distributed to vehicles under pressure (in the range 200-220 kg/cm^2, 2800-3100 psi) to allow the vehicle tanks to contain an amount of gas capable of guaranteeing a sufficient vehicle autonomy.

As distribution networks must meet users' needs, stations may be classified as two kinds: one serving public and one private vehicles. In the case of commuter buses and coaches, filling stations positioned along vehicle runs are quite sufficient. On the other hand, privately owned vehicles require a greater number of refueling stations evenly spread over the whole territory.

Two types of refueling systems are employed:

1) Compression stations, supplied from local methane pipelines, and

2) Filling stations, supplied by cylinder trucks where no local methane pipelines exist. (Cylinder trucks are filled at compression stations also equipped for this purpose).

Two important factors to be considered in station siting are having pipelines feed at great enough pressure to reduce energy absorption due to gas compression, and at the same time be as close as possible to fuel demands.

The major factor in designing refueling stations is their nominal potential. The size of the territory to be served must be taken into account so that vehicles are not compelled to travel overly long distances to reach a station. For this reason, the best solution is to locate stations in well-trafficked centers and spread them conveniently throughout the territory.

The refueling station includes the following equipment:

- **Flow Metering Equipment** , which allows metering the natural gas feeding the station.
- **Compression Units** , which supply compressed gas to storage, allowing it to be maintained at storage pressure.
- **Storage Facilities** , for the compressed gas.
- **Gas Dispensing Equipment** , to fill vehicle tanks.
- **Infrastructure** , including service areas for station staff and customers.
- **Utilities** , to meet acclimatization, lighting, sanitary, and safety requirements.

Such a CNG refueling station, one which supplies vehicles from the CNG storage facilities, allows a cut in the overall number of compression units and their auxiliaries. In such stations the compressor number depends solely on the station output, without being dependent on the quantity of vehicles to be served, as is the case where each vehicle tank is directly filled by a single compression unit. As compressors constitute one of the major installation outlays, fewer compressors means saving in terms of capital costs. Additional savings result as time goes by since compressors, being the station's only dynamic components, are, like all machines, subject to wear and must be routinely maintained. Thus a minimal number of compressors means minimal overall costs.

Basic Compressor Features

In CNG refueling stations, the compression units compress the gas from the supply network to the station storage facilities.

Compressor operation, i.e. start-stop, is automatically controlled to maintain the storage supply pressure within a suitable range. Compressors are equipped with safety and remote-control devices to avoid the constant presence of the station attendants who, as a rule, only provide vehicle servicing. Although the compressors are relatively small-sized (60-120 Kw 80-160 HP), they perform rather heavy duty service because of the high pressure involved. Therefore since regular station personnel have no special training in compressor operation and maintenance, these machines must allow prolonged and safe operation of all components. Moreover, the machines must be of extremely simple configuration and easy to maintain.

The most widely used compression units in CNG refueling stations are reciprocating compressors. There are two, three, and four-stage machines (See Figure 2) to cover all ranges of station supply pressures, which may vary considerably depending upon source (urban gas distribution network or high pressure pipeline).

Safety

In Italy, construction of refueling stations and vehicle inspections comes under the control of domestic authorities who issued safety standards in the early 1950's. These can be found in the Italian government standard "Fire Prevention in Methane Pipelines and Refueling Stations."

Upon installation, the natural gas supply system is inspected periodically by the competent authorities; e.g. supply tanks are inspected during manufacture and thereafter every five years. Also, the maximum life of vehicle tanks cannot, by law, exceed 30 years.

Safety experts and users of natural gas-propelled vehicles rate current Italian safety standards covering vehicle equipment as up-to-date. Yet, on the basis of the long experience accumulated in Italy, construction standards of fueling stations have been assessed as overly restrictive. In fact, requests for loosening up existing legislation are increasingly being aired. In the case of recently built stations, authorities have already expressed approval for slight modifications to existing legislation.

It is our feeling that some of the infrastructure imposed by domestic regulations (e.g. the requirement that reinforced concrete be used all around the storage tanks and as stall dividers) are far too restrictive and may be termed overly cautious. This is confirmed by the fact that no accidents in the pneumatic section justifying use of such protection have ever been recorded in Italian refueling stations. Appendix I of this chapter is a code listing the author's company's proposals for safety requirements for automotive natural gas refueling stations. It includes developments stemming from long-term experience distributing CNG in Italy. However, as this standard is conceived for filling stations to be built in Italy, where official regulation is still very restrictive, it fails to include all of the simplifications the author considers feasible.

On the other hand, safety and technical considerations impose some regulation of the vehicle tank filling procedures, i.e. that tanks be filled directly from storage facilities. This solution makes it possible to:

- Avoid subjecting the vehicle fuel tanks to fatigue stresses, induced by pressure pulsations coming from compression units of the reciprocating type.
- Minimize temperature increase in the vehicle fuel tanks during filling, thereby cutting down refueling times.
- Prevent any traces of liquids infiltrated into the gas during compression (i.e. water and/or oil, which for any reason failed to be separated in the separators) from reaching the vehicle tanks where it is almost impossible to remove them. The storage facilities, on the other hand, are designed to allow the accumulation of left-over droplets that can be drained off by drain valves.

Dispensed gas metering

For the time being, no special devices for metering gas dispensed to vehicles are available. (Amounts are now figured on the basis of the capacity of the tanks to be filled and the gas pressure inside them at the outset and end of the filling operation). Availability of such a device, however, would prove a great all-round station saving, and permit the setting up of self-service stations. Consequently, much attention is currently being expended on rapid development of these devices.

Acknowledgements

The author wishes to thank Nuovo Pignone S.p.A. (Florence, Italy) for permission to publish information regarding company technology on CNG refueling stations.

Appendix I

Safety Requirements for Automotive Natural Gas Refueling Stations

Foreword

The safety criteria set forth herein refer to the units commonly designated "automotive natural gas refueling stations" in which operations relating to filling fuel tanks with natural gas are conducted.

1. Siting

All units belonging to the station shall be enclosed inside wire fencing in accordance with the provisions of Paragraph 2.1 (Construction Features).
No unit, with the sole exception of those used for ancillary services, may be located at less than 10 meters from the edge of the area on which the station stands.
A safety distance between operating units and off-site buildings of at least 30 meters must be observed. This distance is to be doubled when any of the off-site buildings are:

- Schools, public meeting places such as theatres, hotels, hospitals, army posts, shops, marketplaces, railway and bus stations, or the like.
- Of cultural or artistic importance, i.e. museums, galleries, etc.
- Factories or areas where flammable or explosive substances are made and/or handled.

2. Construction Features

2.1 Wire Fencing Enclosure

The station site shall be enclosed inside wire fencing at least 2 meters high and well anchored to the ground.

2.2 Metering Enclosure and Compressor Hall

All gas metering devices and compression units shall be installed in special fireproof enclosures. Proper aeration to prevent air-gas mixtures from forming inside, as well as special gas detectors that activate safety devices, shall be provided.

Furthermore, when hostile environments are encountered, it is necessary to climatize the interiors to a minimum temperature of 5°C.
The room shall be big enough to allow easy machine maintenance, access, and proper positioning and marking of safety exits.

2.3 Storage Facilities for Compressed Gas

Storage tanks may either be installed horizontally, subhorizontally, at ground level, or underground as long as they are in easily accessible and inspectionable positions.
For safety purposes, the storage tank enclosure shall be made of suitably thick reinforced concrete and the enclosure walls shall extend at least 2 meters above the top of the storage tanks.

A system to stabilize the temperature of the gas inside the storage tanks and gas detectors that activate safety devices shall be installed along with the tanks. In the case of hostile environment conditions, everything shall be in turn contained inside a special building, whose structure shall be separate from the storage tank enclosure, and which has the same construction features as the metering enclosure and compressor hall.

2.4 Vehicle Fuel Dispenser Stalls

Adjacent stalls shall be separated by reinforced concrete walls parallel to the vehicle axis. The lay-out, length, and height of the stalls are worked out in relation to vehicle types and how their gas tanks are mounted.

2.5 Facilities for Ancillary Services

These structures shall be fire resistant and fireproof.

2.6 Electrical Substation and Climatization Facilities Building

These shall be built out of fire resistant and fireproof materials and comply with all safety regulations applying to such equipment.

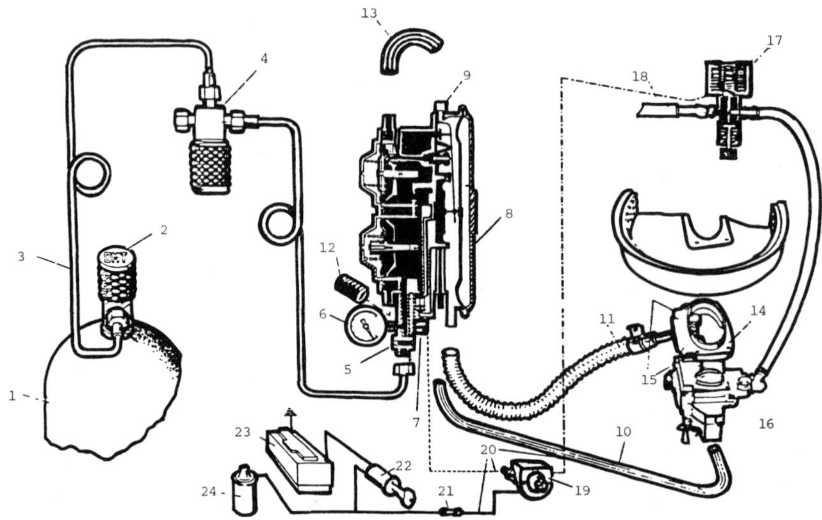

FIGURE 1. Equipment for Natural Gas Fueled Vehicles

1. Methane Tanks
2. Tank Spigots
3. Main Methane Piping
4. Fill-up Valve
5. Purifier
6. Pressure Gage
7. Methane Solenoid Valve
8. Pressure Reducer
9. Minimum Setter
10. Depression-Suction Engine Connection Pipe
11. Mixer-Reducer Connection Hose
12. Water Circulation Pipe (to reducer)
13. Water Circulation Pipe (from reducer)
14. Methane Mixer
15. Maximum Setter
16. Carburetor
17. Gasoline Solenoid Valve
18. Connection Pipe to Gasoline Pump
19. Rotary Switch for Gasoline-Methane Change-over
20. Electrical Connections
21. Fuse
22. Dashboard Key
23. Battery
24. Coil

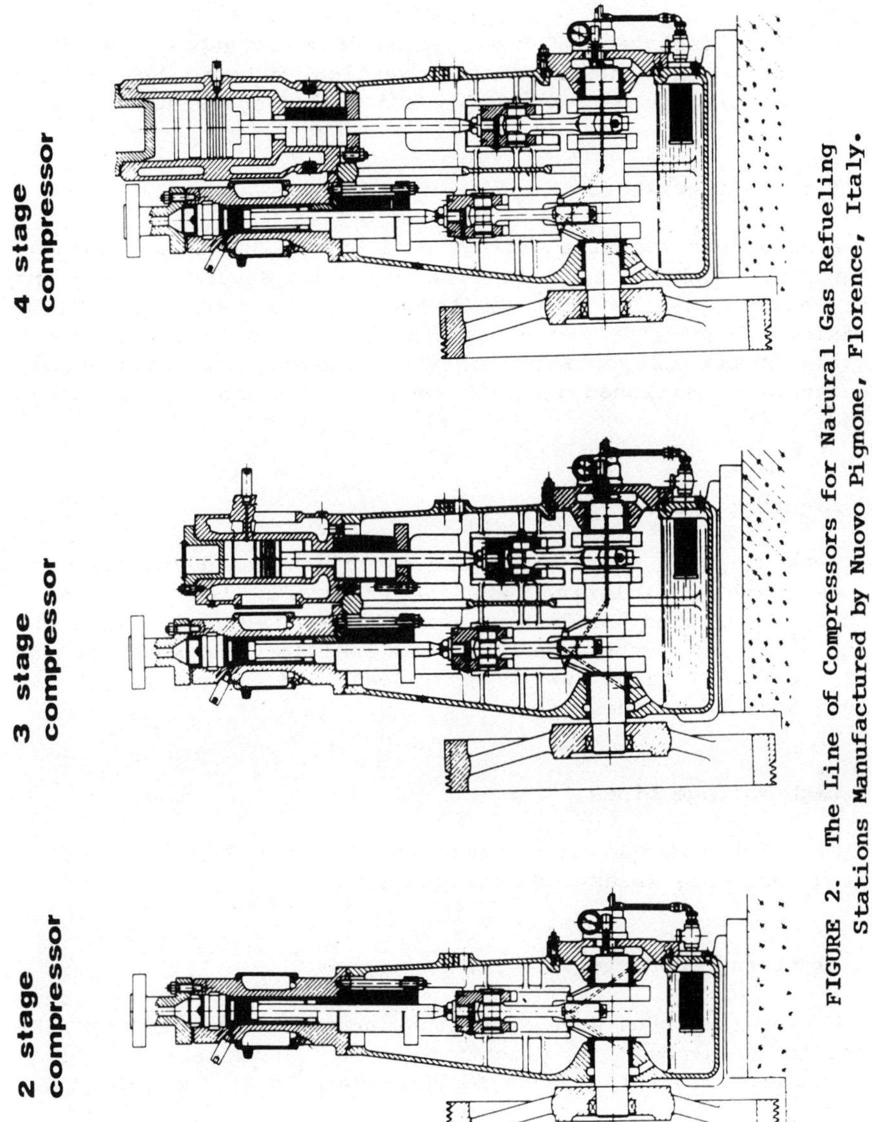

FIGURE 2. The Line of Compressors for Natural Gas Refueling Stations Manufactured by Nuovo Pignone, Florence, Italy.

3. Gas Circuit Safety Features

The gas circuit shall be equipped with a system for venting gas to the atmosphere that is connected with a safe zone. The system must ensure:

- Total isolation of the station from the supply line.
- Sectioning and individual venting of each branch of the gas circuit between supply and fuel dispensing.

4. Fire Extinguishers

A hand-operated fire extinguisher system shall be provided for the storage tank unit(s) and the fuel dispensers. Its control panel shall be set up in a sheltered place as easily reachable as possible. In addition, the station shall also be equipped with a suitable number of portable and wheelable CO_2 and power-type extinguishers, positioned so that each installation may be speedily served.

5. Lighting Plant

Regardless of whatever is set forth in commonly applied regulations, highly-insulating cabling, preferably lead clad, or galvanized steel clad conductors, shall be used for the lighting plant, except for those serving the ancillary services or public facilities. All lamps, switches, valves, and control and/or safety devices must be weatherproof.

6. High Voltage Lines

All high voltage lines shall be at least 20 meters from the station enclosure (measured horizontally).

7. Lightning Conductors

Whenever station size, position, or topographic set-up make it particularly exposed to lightning, the installations shall be equipped with lightning conductors, preferably of the cage type.

INNOVATIVE APPROACHES TO CNG VEHICLES

John E. Wright

General Manager
Gas Service Energy Corporation
Kansas City, Missouri

Our company, the Gas Service Energy Corporation, occupies a very unique position in the natural gas powered vehicle industry because we, as a company, have been operating natural gas vehicles since 1969. We began our pioneering in the CNG field by purchasing six kits from one manufacturer and placing the system into operation in 1970. We have continued to add to our fleet until, at present, there are approximately 700 vehicles in our parent company's fleet operating with natural gas.

Gas Service Energy Corporation is a wholly owned subsidiary of a natural gas distributing company in Kansas City, Missouri,called the Gas Service Company. It is a natural gas distributing company that buys natural gas at the edge of a city and sells it within the city limits. Their gas is sold to 820,000 customers in four states: Missouri, Kansas, Oklahoma and Nebraska.

The gas sales occur in 400 cities throughout the area. The parent company has 2,600 employees and 1,400 vehicles in their fleet. We are targeting about 1,200 of the 1,400 vehicles for conversion to natural gas fuel.

Phase I of our program started with six kits, two more kits were added from a different manufacturer, and then we discovered Dual Fuel Systems of California. At that time, I believe, the company was called Pacific Lighting. Approximately one-third of our fleet is now running on Dual Fuel equipment and, I might say, it is very good equipment. We have operated on it for a number of years and it has provided us with very good service.

When we decided to get into the business of providing

equipment for vehicles, our approach was to assemble all of the mechanics, in our four-state area, who operate and maintain our natural gas vehicles. We asked them to start with the cylinders in the vehicle, go clear through the vehicle system and the refueling system, and tell us all the problems they encountered. We also asked for any improvements they would suggest to be made in the system.

Not only did we get the mechanics' input, but we asked for one additional thing that has really paid big dividends for us: we requested that the local fire prevention chiefs in Kansas City, Missouri and Kansas City, Kansas, visit our system and evaluate the design criteria being formulated. They made several suggestions that were added to our design.

I am going to describe the technical design that our company has incorporated into the CNG systems we sell.

All natural gas fueled vehicles operate in very much the same manner in that they generally have fuel cylinders in the rear of the vehicle. Some companies provide a quarter-turn, high-pressure valve at the cylinders and some do not but, in general, all companies provide shut-off valves either on the cylinders or in the fuel line.

The fuel moves from the filling position to the cylinders through a high-pressure fuel line. The fueling position can occupy many different locations on a vehicle. Vehicle cylinders are pressurized to different levels by different companies. For example, the particular system we sell is pressurized to 2,400 psi.

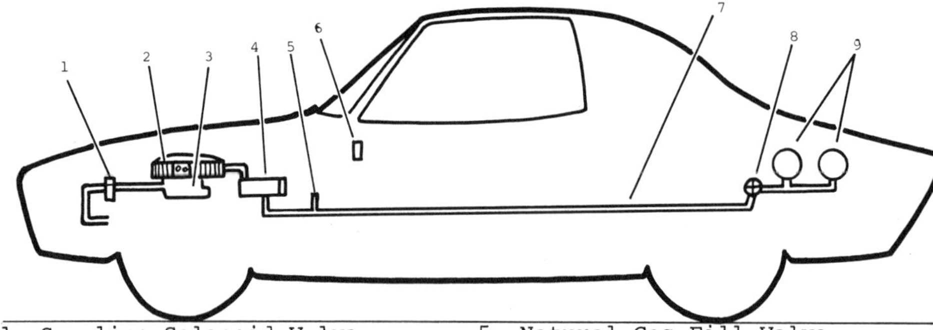

1. Gasoline Solenoid Valve
2. Natural Gas Mixer
3. Original Equipment Gasoline Carburetor
4. Pressure Reducer and Natural Gas Solenoid Valve
5. Natural Gas Fill-Valve
6. Fuel Selector Switch and Gauge
7. High Pressure Fuel Line
8. Manuel Shut-off Valve
9. Natural Gas Cylinder(s)

FIGURE 1. Components of a CNG Vehicle System

The systems in Italy that I saw last year are pressurized to 3,000 psi.

The cylinders that we and many other companies use are Department of Transport (DOT) approved. They are rated at 2,400 psi and, at the time of manufacture, are pressure tested to five-thirds times their rated working pressure which, in the case of a 2,400 psi natural gas cylinder, means that 4,000 psi of water pressure is used to test the strength of the cylinder at the time of manufacture.

Federal regulations require that each cylinder should be pressure tested at the 4,000 psi test pressure level every five years thereafter. We think that this is a very good plan. We, as a company, have operated vehicles for 11 years and sometimes trade vehicles off in less than a five-year period. We have the cylinders tested and recertified before we place them in a second vehicle. You can see that there is a tremendous safety factor built into the cylinders.

There is also a tremendous safety factor built into the high-pressure fuel line. Once the tanks are recharged, the fuel is provided to the engine and the vehicle is started. The fuel leaves the cylinders, passes through the fuel line and then to the regulator package located under the hood. Depending upon which system is utilized, there may be two or three pressure levels that the gas passes through. In the case of our Dual Fuel System kits, the gas pressure is stored at 2,400 psi, drops to 55 psi and then regulates to about two inches of water column pressure. The system feeds this slight positive pressure into the natural gas mixer that is mounted on top of the gasoline carburetor. A set of fuel shut-off valves is provided so that you do not use two fuels at the same time in the engine. These shut-off valves consist of a solenoid valve in the natural gas line and a solenoid valve in the gasoline line. A selector switch is on the dashboard that selects which fuel you will use. All systems are similar except for different regulator packages. Some manufacturers will have different pressure levels than others. Our system has three pressure levels that regulate the pressure from 2,400 psi to atmospheric pressure.

Refueling systems are generally set up so that a high-pressure compressor compresses gas at pressures up to 3,600 psi. The gas is stored in high-pressure DOT 3,600 psi cylinders. The cylinders have suitable valving arrangements so that gas can be placed on board a vehicle in less that five minutes. Generally speaking, most systems require approximately 3 1/2 to 5 minutes to fill a two cylinder vehicle.

In addition, if you have the desire to slow-fill vehicles, you

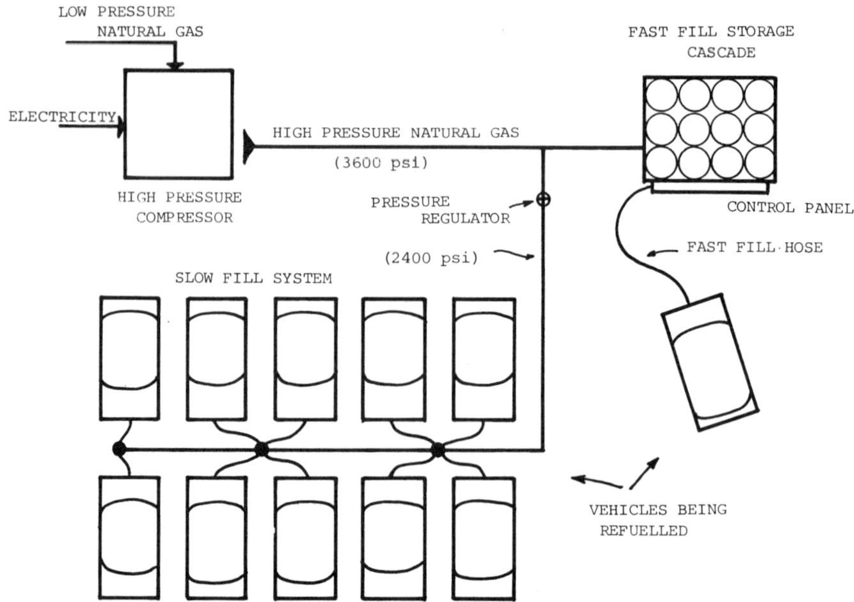

FIGURE 2. Components of a CNG Refueling System

can utilize a pressure regulator in the system and fill any number of vehicles to 2,400 psi on an overnight basis. Generally speaking, you design your overnight refueling system so that the compressor is running during the entire overnight period. In our particular company we fill from about 4:30 at night to about 7:00 in the morning, at which time the vehicles are disconnected. With this arrangement you have both fast-fill and slow-fill. Systems can also be designed for fast-fill only or slow-fill only.

Naturally, when you add the fast-fill capability to your system, you increase the cost dramatically because it costs a great deal of money to contain the gas at 3,600 psi.

A typical compressor cabinet is a weatherproof housing with all the gauges mounted behind a lexan cover. The electrical wiring is explosion proof and intrinsically safe. In case of a leak or an electrical spark, the design is such as to preclude ignition of the natural gas-and-air mixture.

The electrical system on the compressor is designed to Class I Division I, Group D specifications. The cabinet is designed so that it can easily be set in place with a forklift. Two utility connections are required for the unit: electricity is connected to the control panel and, on the back side of the unit, a natural gas connection is needed. Two additional connections are required.

One is a high pressure gas outlet and the other one is a condensate drain outlet.

Compressors are provided in different sizes. Our compressor sizes are 3 1/2 cfm, 7 cfm, 17 cfm and 50 cfm. Roughly speaking, they will handle something in the neighbourhood of about eight vehicles a day for the smallest unit. The 7 cfm unit will handle 16 vehicles. The 17 cfm unit will handle approximately 38 vehicles a day and the 50 cfm unit will handle about 85 vehicles a day.

CNG is stored in a cascade at 3,600 psi, which contains 12 high-pressure cylinders of 450 cubic feet per cylinder for a total storage capacity of 5,400 cubic feet.

The dispensing control panel is mounted on the cascade and is built as a unit at the factory. The unit can be uncrated, set in place, and only requires two lines: the first is the high-pressure gas connection from the compressor and the second is the condensate drain line to the mast. The mast is ten feet tall and is provided with a weather-proof cap. The mast is used to vent any gas that is vented from the vehicles after filling or any other system gas such as condensate drain gas that is vented. The venting concept was suggested by the local fire chiefs in our area and we like the concept.

The cylinder rack is very unique in its design. Note that the cylinders are sitting in a very sturdy metal frame. The frame is designed with a hinged gate that can be unbolted and moved aside so that the cylinders can be easily removed every five years for retest. The plumbing system is also designed to be easily disconnected and, with the swinging gate, you can easily remove the cylinders from the side of the cascade rack.

One of the problems we have encountered in operating vehicles for the last 11 years is that, periodically, we will have someone drive away with a vehicle still connected to the refueling system. We have ignition disarming switches installed in our vehicles and in theory it is an excellent idea. In practice, however, you are at the mercy of the mechanic and his workmanship in installing the last switch on the vehicle.

In our case, we have experienced some instances where the switches were not functioning and the vehicles actually were driven or rolled away while they were hooked up to a refueling hose. We have experienced this at one of our plants approximately ten times. Now we have designed a fill post so that when vehicles drive or roll away, the hose is automatically disconnected from the fill post and there is no escaping gas or damaged piping system.

Our fueling stations have one-position, two-position and

four-position posts. We now have a design that will disconnect anywhere in a 360° arc around the post instead of approximately 30°. If you pull on a hose anywhere in the 360° arc, it will automatically disconnect from the piping. There is no escaping gas because male and female check valves are built inside the hose quick-disconnect coupling and these seal off any escaping gas when the hose disconnects.

Good installation practice for cylinders inside a CNG vehicle dictates that a sealing kit be installed over the ends of the cylinders where the piping is located. Any leakage that might develop in cylinder piping is contained by the sealing kit and vents through the floor on the vehicle. We think that this is a mandatory item to have on any inside cylinder installation. Our 11 years of operating systems have shown us that, periodically, we will encounter a slow leak in a system. These small leaks will be in the cylinder rupture disc or valve threads.

We have experienced systems where the rupture disc has leaked after being in service without leakage for several years. The disc is tightened to proper torque and the gas pressure pushing against them for a number of years slightly extrudes the 212° alloy that backs the disc. This slight extrusion will then allow minor (non-hazardous) leakage which is detected by the odorant in the gas. Interestingly enough, some of the earlier vehicle systems that we placed in service in 1970 are still in operation today. The vehicle kits have outlasted two previous vehicles and are now on a third vehicle.

Let me explain the function of the rupture disc. Fuel cylinders are extremely rugged and are hydrostatically tested to 4,000 psi at time of manufacture. The burst strength is approximately 7,000 or 8,000 psi so they are very conservative in their design. If a cylinder should be in a fire, however, there is no telling how high that pressure could go, so a device is needed to relieve the pressure. The rupture disc is very ingenious in its design. The Department of Transportation in the United States (formerly the old Interstate Commerce Commission) has regulations insisting that every cylinder that contains gas shall have a rupture disc installed. If the temperature gets too hot around the cylinder, the soldered backing that backs the rupture disc will melt and allow the pressure of the gas in the tank to push on the rupture disc and cause it to rupture at the designed pressure. When the disc ruptures, the four exit ports on the rupture disc housing will release the tank contents in a controlled manner, rather than in the catastrophic manner which would occur if the cylinder slit from end to end.

I have discussed, with wholesale cylinder dealers in Kansas City, some of the fires that have occurred in cylinder plants around nitrogen, oxygen and other cylinders, and they say that it is really a dramatic thing to see the results of the controlled exiting of gas handled by rupture discs.

We have installed systems on pickup trucks, police vehicles, and large newspaper delivery trucks. Such systems can have the cylinders hanging underneath the bed of the vehicle. We have discussed the short-range limitations of vehicles utilizing natural gas. We do not, however, really look upon natural gas vehicles as being short-range vehicles. We consider them to be extended-range vehicles because, in addition to the gasoline supply that you formerly enjoyed, you have added three gallons for every cylinder that you install on the vehicle. Therefore, any car that I have ever driven on natural gas (and I am presently driving my third one) will go on for longer than I can drive on the highway.

One example I can cite is a dump truck that drives over 100 miles a day in a very limited geographic area. There are four cylinders on this particular vehicle, each one holding the equivalent of three gallons of gasoline. It is filled twice a day and the truck runs five days a week, 52 weeks a year. It uses approximately 6,200 gallons of gasoline equivalent in natural gas fuel a year. At our cost differential in Kansas City, Missouri, of 72¢ a gallon, the truck saves our company $4,300 a year in operating costs.

What is even more dramatic for the gas industry is that the truck <u>uses as much fuel a year as six houses.</u> This is why our parent company entered the CNG vehicle business. Gas Service Company has 820,000 customers in their service area and we have seen our residential and small commerical gas loads drop from 210,000 cubic feet per customer in 1970 to 169,000 cubic feet per customer in 1980. The comparable new customers coming on-line are now using 125,000 cubic feet per year, so you can see that we have gone from 210,000 cubic feet average to 125,000 cubic feet average. This creates a cash flow problem and we make up the difference in our revenue by raising our rates. When we do that our customers are not pleased, so the CNG vehicle concept is an attractive alternative.

Interestingly enough, when we decided to get into the marketing of these systems we decided to do it just in our general service territory of Missouri, Kansas, Oklahoma and Nebraska. We are now worldwide.

The average fleet vehicle in our area uses less than five gallons of fuel a day. If you consider five gallons of fuel a day times five days per week, times 52 weeks per year, that is 1,300 gallons per year of fuel going through a five-gallon per day vehicle. This is equivalent to one and a quarter houses in our service territory. Note the fact that the average vehicle travels less than 50 miles a day - in our fleet system of 1,400 vehicles we average from 30 to 50 miles a day for each vehicle.

Our vehicles need to carry less than one day's fuel supply. Our fleet vehicles - from the supervisor's car to the service vehicles-report to the same location every day where they are refueled.

Consider the fact that the average 25 car fleet will use as much gas in a year's time as 32 houses. Every house that we add to our natural gas distribution system costs an average of $1,300 per home. This means we spend over $40,000 to connect 32 houses to our gas distribution systems. For each 25 car fleet we only have to provide one service line and one meter at a cost of $2,000. The customer owns the CNG equipment.

It is obviously a tremendous marketing tool for gas companies.

Another example is a welding truck that we use for welding gas pipelines. Not only does the engine on the vehicle run on natural gas but so does the electric welder. We suggest that you convert one sometime and let your welders try it. Every welder that we have converted runs better than it does with gasoline. We are also operating air compressors on our service trucks with CNG.

I have had people say, "Why carry all that weight around in your car?" I tell them, "That's my favorite uncle. He weighs about 250 pounds and as a result of hauling him around in my trunk, I can save 70¢ a gallon in fuel costs. So what if I have to put a set of helper springs on the car?" (These cost about $50.00 and are easy to install.)

Personally, I drive a 1977 Chevy Impala that now has been driven 78,000 miles. The servicing of this car has been minimal. It was first tuned up at 32,000 miles and again at 67,000 miles. The manual says to tune the vehicle at approximately every 12,000 miles. Before we tuned it up at 67,000 miles, we used a Sun Computer II machine to check the ignition system and we found two spark plugs were breaking down slightly, so all of them were replaced. The oil in my car is changed every 6,000 miles, but that is not because the oil really needs it since it is still very clean. After all I drive 95% of my mileage on natural gas. However, I was part of the "old school" which dictated that I should change my oil every 2,000 miles. Note that I am now up to three times that interval.

As a general rule of thumb, for vehicles that we operate on natural gas in our parent company's fleet, we have extended the time between tune-ups and between oil changes by at least 50% more

FIGURE 3. This dump truck, which runs over 100 miles a day, saves $4,300 per year in operating costs

FIGURE 4. Fuel Gauge and Selector Switch

than the book calls for.

Our first system in 1970 had numerous first stage regulator problems, such as freezing, relieving natural gas out through the relief valves, poor seating, etc. We redesigned the system and used more reliable equipment that would hold together and give good performance on the vehicle.

Our present system utilizes the Tartarini regulator and gasoline solenoid valve which are made in Bologna, Italy. We use an American-made carburetor and Italian-made fuel gauge and fuel selector switch. The system drops the pressure in three stages. Stage I drops 2,400 psi to 40 psi, Stage II drops 40 psi down to about 10 psi, and Stage III drops from 10 psi down to atmospheric pressure. This is a very unique system, as you cannot provide gas from the regulator unless there is physical suction on the end of the fuel hose from engine combustion air.

There are three things that have to happen inside the regulator before gas will be delivered to the vehicle engine. Firstly, you have to turn on the 12-volt electric solenoid valve. Secondly, you have to crank the engine to establish a manifold vacuum which is used to arm the internal mechanisms of the regulator package. Thirdly, engine combustion air has to physically suck on the end of the fuel hose to pull gas out of the regulator. This concept works very well in our vehicles.

One of the reasons we have adopted our present gasoline solenoid valve is that, if the solenoid coil burns out, a little grey screw can be unscrewed to open a built-in bypass for the gasoline system.

The fuel gauge and fuel selector switch operate so that, when you are on natural gas, the switch is in the twelve o'clock position. When you want to switch to gasoline it is turned to fill carburetor (FC) at the three o'clock position. This turns on the gasoline supply at the same time you are running the engine on natural gas. When the engine starts to flood out, because it is full of gasoline in the carburetor as well as natural gas, you switch down to the six o'clock position which turns off the natural gas source. The vehicle is then driven on gasoline until you return to the fueling station.

We advise all people to whom we sell the system not to remove the vehicle's gasoline tank. Our parent company will not remove the gasoline tank for the following reason: We sell gas in 400 communities and, if we have a natural catastrophe in one of those towns - like a fire, flood or tornado - and the natural gas system is out of action, the customers aren't using gas, which is our life blood. We then mobilize employees from all over our system and send them to the area to get the system back in operation just as quickly as possible. Customers learn to depend on our natural gas and we want them to have the best possible service. We are called the Gas Service Company because we supply "gas" and we provide "service" to our customers. When they are out of gas, they are out of our service and we want to keep them operating.

When we mobilize vehicles and send them to repair damage it may be to a town where we do not have a CNG system or their CNG system may be just big enough to supply the vehicles in that system.

The fuel gauge shown operates on a very unique principle in that it converts pressure to electricity and gives an indication of fuel contents. Our CNG vehicle drivers wonder why there is more pressure on the gauge in the summer than in the winter. It looks that way on the gauge because our systems operate with dome regulators that regulate the vehicle filling pressure to different levels in summer and winter. Imagine filling a cylinder to 2,400 psi and closing the valve. If we put the cylinder in a hot room and heat the cylinder to 100°F, the pressure in that tank is going to increase. It will increase on a 100°F day to approximately 2,700 psi. If we take the same cylinder and move it to a refrigerated room where the temperature is 0° F, the pressure will decrease to approximately 1,700 psi. By our system the same amount of fuel is provided summer and winter.

Six months ago we decided that we were going to redesign the fuel gauge because each time we had to go through a long explanation as to why the pressure gauge and fuel contents were O.K., even though they appeared to have more fuel in the summer than in the winter. We have now developed a fuel gauge that is totally unlike anything in the world. It is a pressure-temperature compensated fuel gauge. We read, at the fuel line, a pressure which is fed to a chip circuit on the back of the gauge. We also have a temperature-sensing device located at the cylinders that sends a signal to the gauge circuit. The circuit correlates the two items of pressure and temperature and indicates the tank contents. Note that it doesn't tell pressure. We think that this gauge has widespread utility appeal for anybody that installs CNG vehicle systems.

The face of the gauge contains a green lightbulb which indicates natural gas usage. We also have an amber lightbulb that indicates gasoline operation.

In the old days, not all vehicles ran well on natural gas. One of the things that we experienced periodically, back in the early days of CNG vehicles, was driver reluctance. The drivers complained about poor performance of the vehicles and they were

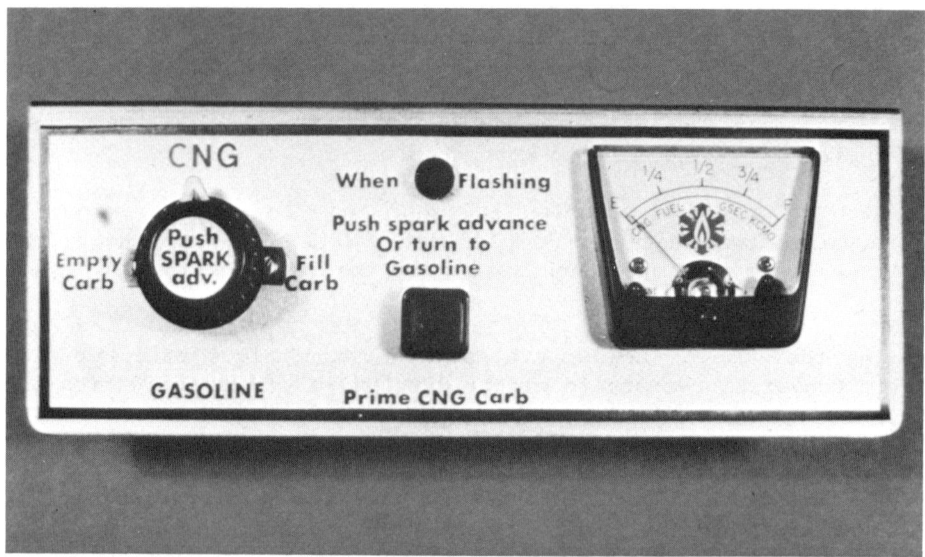

FIGURE 5. Pressure-Temperature Compensated Fuel Gauge Developed by the Gas Service Energy Corporation

FIGURE 6. The vent-hose and fill-hose plug into the front of some vehicles

right. This was because we were not doing anything with the ignition timing other than advancing the distributor from four to six degrees. We are now installing spark advance devices on vehicles. We are using the autotronic system because it is readily available for most vehicles. The improvement in performance removed the objection that operators expressed regarding low power of the vehicles.

In the early days we encountered vehicle operators that complained about low power and investigation revealed that they compared the performance of CNG vehicles with their personal cars with lots of power, such as 440 Dodge Magnums. In such comparisons, CNG looked deficient for several reasons. First of all, an accelerator pump is not present on the CNG mixer to waste fuel and secondly, a power loss of at least 10% occurs in even the best of vehicles. This is due to the volumetric efficiency of natural gas displacing gasoline. If anybody suggests that a vehicle runs as well on natural gas as it does on gasoline, they are wrong. But usually you can't feel the 10% power loss.

We have used manually operated filling panels for ten years in our system. We also provide automatic panels so that anyone can easily fill their vehicle in less than five minutes. Everything in the system is built to extremely strong specifications. All components have 5,000 psi working pressure rating. A relief valve is also incorporated into the system so that the compressor cannot overpressure the system.

The pressure switch on the compressor is set to shut off at 3,600 psi and, if it doesn't, a relief valve on that fourth stage of the compressor will relieve pressure at 300 or 400 pounds above the shut off point. We place an additional relief valve in the control panel so that there are three backups in the system: 1) the compressor shut off switch; 2) a relief valve in the compressor, fourth stage, and 3) the relief valve in the control panel.

A hole is placed in the grill and a refueling point is established on each vehicle. On fast-fill vehicles, we put the refueling system underneath the hood since there are not very many people that will drive off with the hood up. A two-hose assembly is used to fill vehicles. The left-hand valve is the high pressure valve that allows high pressure gas to flow from the compressor to the vehicle. We use a different type of refueling connector at the vehicle than the Dual Fuel male-female probe. We use a quick disconnection coupling, because of a suggestion received from the local fire chief's office. They wanted a system designed so that high pressure gas would not vent into the atmosphere when we opened the hose valve. When a vehicle is connected, the vent valve is closed and the fill valve is open. When the vehicle is to be disconnected, the fill valve is closed and the vent valve is opened. This vents all the gas that is in the refueling coupling from the check valve inside the vehicle back to the fill valve and then out through the vent hose. All you hear over your shoulder is a small sound as the gas vents ten feet above you through the vent cap and into the air.

There is a yellow dust cover that fits over the refueling probe when the vehicle is being driven. There is nothing magical about putting the filling position on the front of the vehicle. We have installed fill points on the back or sides of vehicles. The quick disconnect coupling at the horizontal bar is a twin of the one that is on the car refueling point. Note the pivot point below the horizontal bar. If the vehicle pulls away while it is hooked up, it pulls on the ring that the hose passes through. The pivot point then depresses the quick disconnect coupling, and then snap ring and pressure physically force the male connector out of the female connector. The car driver should stop because the sound is loud, like a shot from a pistol. If he continues to back up, he will then break the little plastic elbow connected to the vent hose. If he backs up further, the rope that suspends the hose apparatus on the hose reel contains a weak link. It will break next to the reel and the hose just drags away from the post. We have never had anyone move from the position with the hose because they can feel when they pull it loose and also hear the noise.

Every operator that obtains gas must have a panel key. When the key is inserted and turned on, the next step is to push the green button. Everything else is automatic. As the gas comes out of various cylinder banks, it actuates a green indicator so that you know which bank of cylinders is feeding the gas into the vehicle. The vehicle gauge in the lower right hand corner tells you what pressure is being placed in the car. If somebody forgets to turn the key off, the system will automatically, after a present interval of time, shut itself off. If the operator forgets to push the stop button to shut off the valves going to the hose, the system will automatically shut off when the key is turned off. We want the system to shut off when the filling sequence is completed.

SOME NATIONAL PROGRAMS FOR ALTERNATIVE FUELS

THE NEW ZEALAND EXPERIENCE

Peter J. Graham
Assistant Secretary (Oil & Gas)

Ministry of Energy
Government of New Zealand

The New Zealand Government is actively involved in a program for converting 150,000 motor vehicles, about 10 percent of our total fleet, to CNG by the end of 1985. The program began formally in mid-1979 when the target was announced by the Minister of Energy. In describing, from the Government's viewpoint, progress and problems with our program I am conscious that some aspects are specific to the New Zealand scene. However, I hope there is sufficient wider relevance to make this sharing of our experiences worthwhile.

I would like to put it briefly into a policy context and to give a little history.

New Zealand Energy Situation

Oil supplies 47 percent of New Zealand's primary energy requirements. Apart from a small amount of condensate produced with natural gas, all this oil is imported either as feedstock for refining or as finished product. The cost of this has escalated dramatically both in dollar terms and as a proportion of imports. Because the value of our exports has increased much more slowly than the cost of oil we now have to export three times as much to pay for the same quantity of oil imports. The economic consequences for New Zealand have therefore been severe and the benefits of substitution are great.

There are two unusual features of New Zealand's oil consumption which are revelant to CNG:

- 86 percent is used for transport, and
- of this, petrol comprises 54 percent.

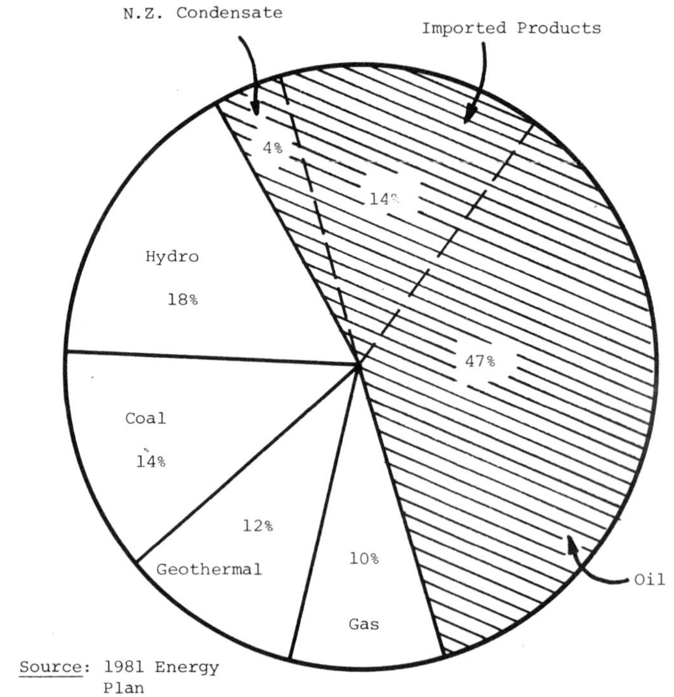

FIGURE 1. New Zealand's Primary Energy Requirements

New Zealand's Energy Resources

One small oilfield has recently been discovered containing 115 petajoules (PJ) of oil and 7 PJ of associated gas. Expected production is about 3000 barrels per day (bpd).

The Kapuni onshore gasfield has 430 PJ of gas and 220 PJ of condensate. The much larger Maui offshore field contains 550 PJ of gas and 780 PJ of condensate. Current gas consumption is 39 PJ.

Coal is New Zealand's largest resource with 57,000 PJ or 4000 million tonnes being used each year at present; coal is our sleeping giant.

There are also considerable remaining resources of hydro and geothermal.

Development of the Maui field, which was discovered in 1969 and started producing in 1979, was originally based on the use of the gas for electricity generation. At that time power demand was increasing at 7 percent per annum.

TABLE 1. NEW ZEALAND'S OIL BILL

	Million tonnes	cif cost ($) (cargo insurance freight)	Percent of all imports
1972	4.04	94	5.1
1973	4.44	115	5.3
1974	4.51	306	8.5
1975	4.23	375	11.9
1976	4.05	483	12.2
1977	3.92	526	13.4
1978	3.74	492	14.5
1979	4.00	725	13.1
1980	4.00	1272	16.9
1981 (estimated)	3.76	1489	
1982 (estimated)	3.85	1600	

Source: Energy Data File

However, the increased value of the gas for oil substitution, a fall in the rate of growth of electricity consumption and greater emphasis on renewable resources (hydro and geothermal) for power generation pointed to the use of the gas to replace oil imports. Figure 2 shows the pattern of consumption of Maui and Kapuni gases now expected up to 2009.

History of CNG in New Zealand

CNG began in New Zealand in 1976 when the Auckland Gas Company converted some of its vehicles and established a small refueling station. In 1977 the New Zealand Energy Research and Development Committee (ERDC), a semi-independent Government-funded body, awarded a research contract to the Wellington Gas Company, for a thorough technical and economic evaluation of CNG in the New Zealand context.

The results were favourable and, fortunately, became available at the time of the Iranian oil crisis in early 1979. The Minister of Energy asked the ERDC to prepare an implementation program. This was completed in about six weeks and four alternative plans, with the actions needed to implement them, were also prepared.

The Liquid Fuels Trust Board, a body funded by a 0.1¢/liter levy on all petrol and diesel sold in New Zealand and charged with investigating methods of reducing New Zealand's use of imported oil, funded a team to carry out an in-depth study of Italy's CNG system.

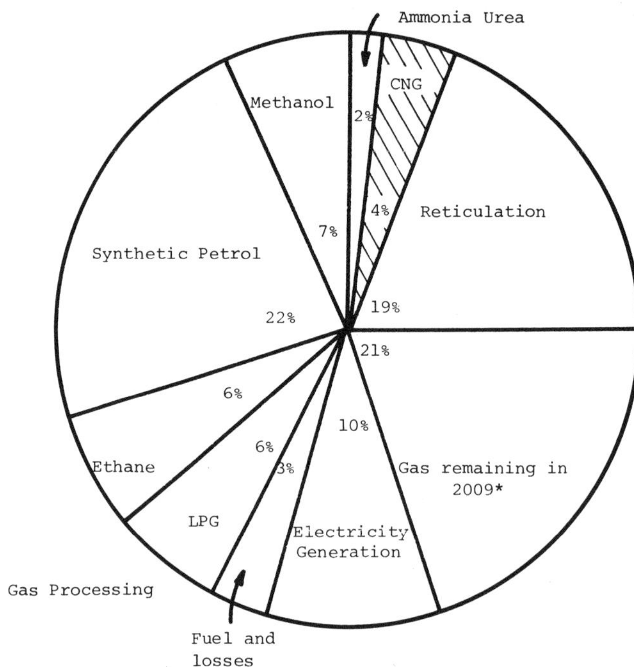

* could be used by a new project before 2009 if the second Maui platform was built

Source: 1981 Energy Plan

FIGURE 2. Consumption of Maui and Kapuni Gas until 2009

Implementation of the CNG Program

The complexities and uncertainties identified by the implementation study did not, rather surprisingly, create panic and in July, 1979, the government announced a target of 150,000 vehicles to be converted to CNG by the end of 1985. At the same time it introduced incentives for the conversion of business vehicles and the establishment of refueling stations.

The basis of the 150,000 vehicles is shown in Table 3.

The number of vehicles which it is in the national interest to convert is higher than the number for which it is financially viable mainly because the reticulation system is regarded as a sunk cost in national resource terms and results, therefore, in a lower gas cost. This, of course, provided the Government with the rationale for providing incentives.

TABLE 2. ANNUAL VEHICLE CONVERSIONS TO ACHIEVE TARGET

	1979	1980	1981	1982
Annual Number	3,000	10,000	20,000	25,000
Cumulative Number	3,000	13,000	33,000	58,000
Annual Petrol Savings (petajoules)	0.2	0.7	1.8	2.9

	1983	1984	1985
Annual Number	30,000	30,000	30,000
Cumulative Number	88,000	118,000	148,000
Annual Petrol Savings (petajoules)	4.2	5.5	6.9

TABLE 3. VEHICLES FOR CONVERSION

Petrol engined vehicles in the gas reticulated area	750,000
Vehicles physically suitable for conversion (Delete vehicles which are too old, too small, etc.)	600,000
Vehicles for which conversion is financially viable (Delete low mileage vehicles, vehicles travelling frequently outside gas-reticulated area, etc.)	300,000
Vehicles which are nationally viable	420,000
Vehicles for which conversion is acceptable to owner (taking into account loss of boot space, power loss, range reduction, high initial costs, conservatism, etc.)	150,000

In the Iranian crisis New Zealand had carless days and petrolless weekends. CNG was exempt from these restrictions and, with growing concern about security of oil supply, enjoyed a flying start.

Developments in 1980

After the initial burst the rate of conversion dropped off from about March, 1980. The main reasons were the easing of the oil crisis, slow development of appropriate standards, lack of publicity, shortage of training facilities (and therefore trained installers) in some areas, problems over cylinder approvals including the withdrawal from use of the most popular make, few refueling stations and lack of public confidence that the Government's intentions were serious. Factors accounting for this lack of confidence included the continued application of duty and sales tax to conversion kits and the absence of a program for the conversion of the Government's own vehicles.

Utilizing the results of a public attitude survey carried out by the ERDC, discussions with the CNG industry and other information, a package of further measures to encourage CNG was developed. The package was approved by the Government, and announced by the Minister of Energy in November, 1980.

MEASURES FOR ENCOURAGING CONVERSION

1. A $200 grant is offered on each CNG conversion kit to offset duty and sales tax (applies also to kits used with biogas).

2. 100% of the cost of conversion of business vehicles can be written off for tax purposes in the year the expenditure is incurred.

3. A 25% grant for equipment is offered for CNG filling stations.

4. For taxable enterprises the balance of cost of CNG filling station equipment is eligible for an immediate tax write-off.

5. The Development Finance Corporation is willing to consider proposals for financing CNG filling stations.

6. CNG has been removed from price control to ensure market place competition.

7. The present method of collecting road tax on CNG through the Road User Charge System is to be replaced by a fuel tax included in the price of the CNG equivalent to the road tax portion of the tax on petrol.

8. Amounts of \$60,000 in 1980/81 and \$20,000 in 1981/82 and 1982/83 have been allotted for a joint publicity program with the industry.

9. The 100% immediate write-off has been extended to cover assembly-line-fitted conversion kits.

PUBLICITY AND INFORMATION

Ministry of Energy

- Seminars and workshops on CNG.
- Publication of a tabloid "CNG Special" distributed free.
- Publication of a more detailed booklet on CNG.
- Audio-visual on CNG.
- CNG Community Information Folder.
- Hand-out material on specific aspects.

Industry

- Media advertising.
- Sales material.
- Newspaper supplements, TV adverts.
- Gas Association Bulletin.

Joint Industry/Government CNG Publicity Committee

- TV advertising.
- Booklet associated with the TV advertising - in effect a coloured, simplified, shortened version of the tabloid.
- CNG workshop manual - general information on CNG for the installer.
- Media material for local use, e.g. prepared advertisements.
- Posters, signs, booklets, folders.

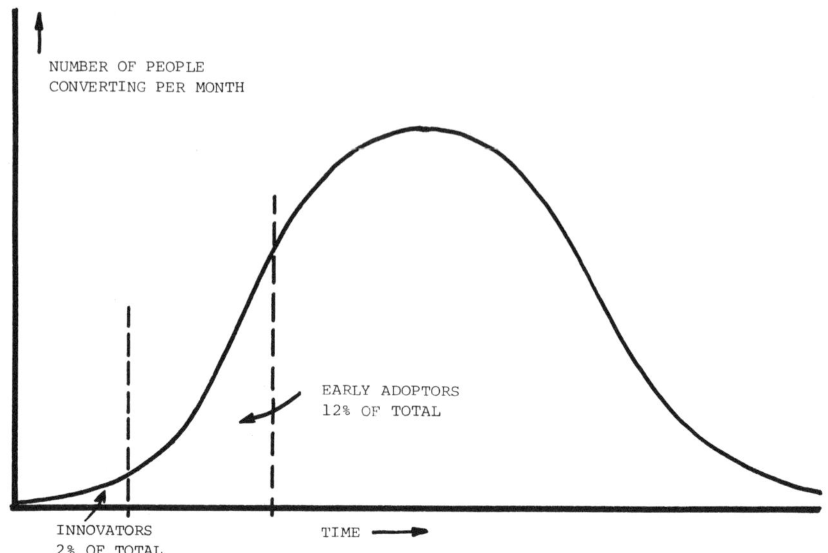

FIGURE 3. Implementation of CNG

Current Status of the Program

There has been a substantial pick-up in kit sales since November last year. However the rate of conversion, at just over 1000 per month, is still below what we would like. Looking at the profile for the introduction of a new technology we are probably now at the "early adopters" phase.

A major problem earlier this year, resulting from the increased rate of conversion, was a shortage of the popular types and sizes of cylinders. More lately there has been a shortage of refueling capacity in some areas. CNG is a classical chicken and egg situation.

Currently there are 46 public refueling stations and 25 private facilities in operation. A further 37 public and 8 private stations are under development and will be open by 1982. These include 20 of the 50 to 60 public stations that Caltex is establishing as explained by Bill Dunning in his chapter. These developments should enable refueling capacity to catch up with conversions and overcome hold-ups at the refueling points.

Two public attitude surveys have been carried out since the November package was announced and the second of these surveys followed greatly increased promotional effort; both showed that there is now very good awareness of CNG and belief in the

Government being serious about its 1985 target. We want to see this public acceptance translated into a faster rate of conversion.

Key Factors

1. National Economics

The national capital outlay for achieving the target of 150,000 vehicles converted to CNG is estimated at $280 million in 1981 dollars. The annual saving on the petrol replaced is $74 million giving an annual return of 26 percent, well above the Government's required minimum of 10%.

The payback on the foreign exchange cost is about 18 months.

2. Financial Return

The average retail price of CNG including tax is 34¢ per liter equivalent of petrol ($1.53 per gallon). There is therefore a saving of 27¢ per liter equivalent. The average cost of a conversion with one bottle is currently $1100. Therefore, for a car doing 10 kilometers per liter (28 mpg) the break-even is 40,000 km (25,000 miles), assuming zero residual value for the conversion equipment. If this equipment is given a value of half its installed cost, the vehicle operator pays off the cost after 20,000 km (12,500 miles) and pockets all the savings from then on.

The break-even distance for a vehicle which does 6.5 km per liter (18 mpg) with two bottles (cost $1600) is a little less.

3. Refueling Station

Several requirements have to be met covering town planning, dangerous goods regulations, and electrical and gas installation inspections. A standard for the design and layout of CNG stations was published last year. A compressor standard has almost been completed. A guide for station developers which steers them through the whole process of setting up a station, from the initial gleam in the eye to opening day, will be published shortly. This was part of a recently completed study on barriers to the establishment of CNG refueling stations. Surprisingly, the study showed no remaining major impediments, although there are still site specific problems. Some of the smaller earlier CNG outlets were stand-alone but almost all are now installed at existing petrol stations. The pattern of distribution of such stations and the availability of gas and economic viability will be the main factors determining the pattern of CNG facilities.

4. Measuring CNG

Currently the pressure difference method is used with a set of prepared tables which allows for different sized bottles, different starting pressures and different ambient temperatures (if a station does not have a temperature correction device such as a dome loader). A programmable calculator with printout is now coming into use and will be required by a standard now being prepared.

One make of on-line CNG meters is commercially available now and two others are under development. These will provide the means of measuring CNG at refueling stations in the future. Minimum requirements for on-line meters are being prepared.

5. Vehicle Conversion

A standard for the conversion of vehicles to CNG and LPG was published last year. Regulations covering the approval and regular checking of conversions should be in operation by 1982. The cylinders must be approved by the Chief Inspector of Dangerous Goods and must be tested every five years. Leasing and hire purchase arrangements for the financing of conversions are available. There are over 300 garages which can carry out conversions.

6. Training

Courses for training mechanics in conversions are provided by technical institutes. To date about 1,500 have attended. Familiarization courses are available for forecourt (pump) attendants and panel-beaters (body-shop operators).

7. Government Fleet Conversion

A program to convert all possible Government cars and vans based in the gas reticulated (pipelined) areas started last year. Currently 1,600 vehicles have been converted. By 1985 this will increase to about 5,000. Not only is this part of the whole conversion effort providing savings for the taxpayer, but it helps to build up public confidence.

8. Gas Supply

This is secure for at least 30 years and more likely 50 years even if there are no further discoveries. Over the next 5 to 6 years over $500 million will be spent in New Zealand on exploration if seismic surveys lead to drilling. The prospects for additional finds of gas, which are more likely in New Zealand's geology than oil, are therefore very promising.

9. Price of CNG

Currently the price of CNG, as I have indicated, is about half that of petrol. The Government has given an undertaking that it will not increase taxes to the disadvantage of CNG relative to petrol. The price of gas, at the source, which makes up about half the price of CNG, is expected to increase at about the same rate as inflation, whereas oil prices will increase faster. The present favourable margin for CNG is therefore expected to be at least maintained.

10. Research & Development

Research capability, providing a center of expertise, needs to be maintained. Work in New Zealand on CNG, carried out particularly at the University of Auckland, is concentrated on conversion kits, and particularly the mixers, the development and evaluation of on-line meters, and CNG/diesel conversion.

11. Trade Associations

A CNG Federation was formed earlier this year.

12. Coordination

A CNG Coordination Committee, set up at the beginning of the program, comprised representatives of the four key Government departments involved, the Energy Research and Development Committee and the Liquid Fuels Trust Board, and two people from the private sector. Unfortunately, the latter had to retire because of difficulties with the rest of the industry. The main function of the committee is to resolve institutional and administrative hang-ups. Since I took over as chairman, in the middle of last year, I have also held periodic meetings with the industry as a whole to maintain liaison.

13. The Government's Role

In summary, it is the Government's policy for the development of CNG in New Zealand to be carried out by the private sector. The Government's role is to provide the right climate in which this can happen - by promoting confidence in CNG, resolving institutional and other impediments, pump priming through incentives, monitoring and funding research.

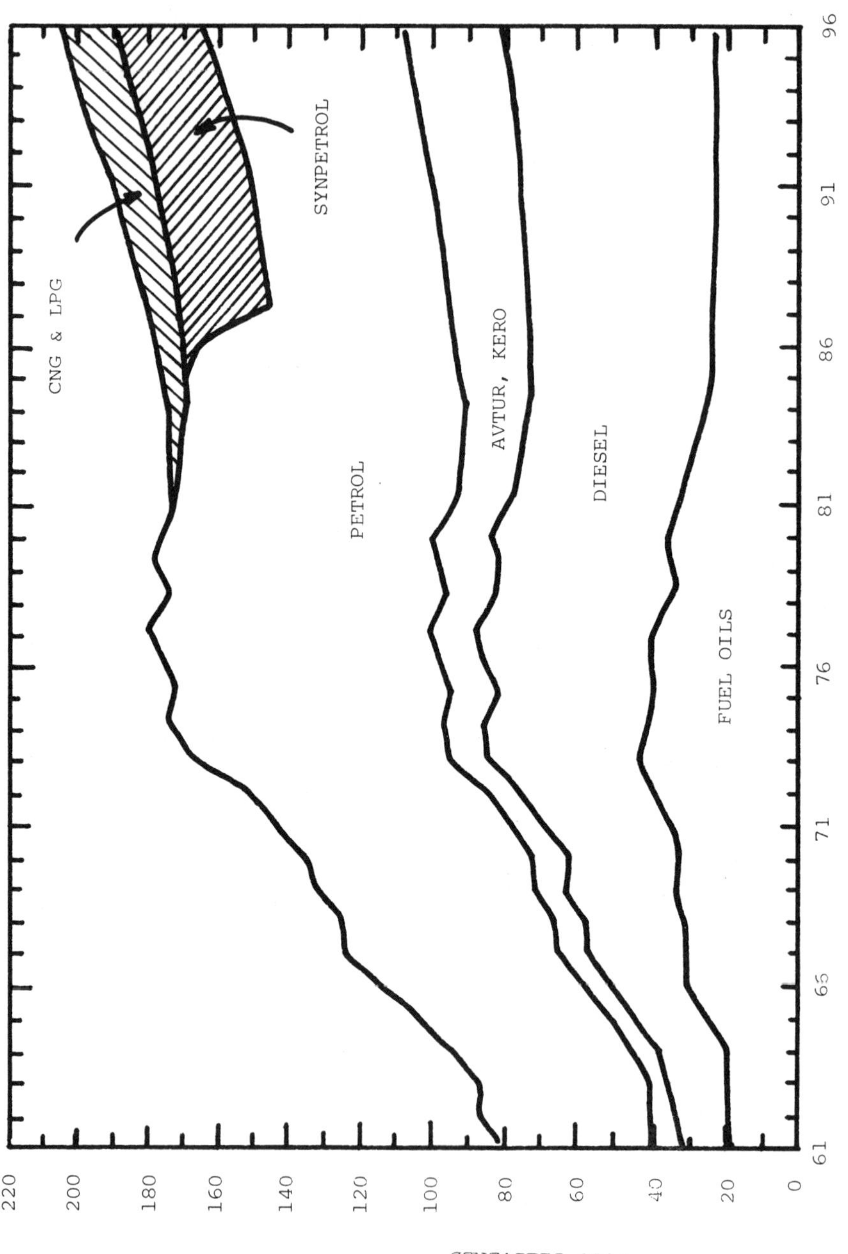

FIGURE 4. Oil Product Demand in New Zealand (1961-1996)

The Future

CNG is the only New Zealand-produced transport fuel which is readily available and financially viable at the present time. Given these factors and our heavy dependence on imported oil, the Government has implemented the action-oriented program I have described.

The main problems with the program which we see ahead are:

1. Maintaining public confidence.
2. Adjusting to changes in the calorific value of natural gas. (At present two different calorific gases are reticulated and used for CNG, soon there will be three. In about seven years all areas should be receiving the same gas. Research is getting underway in this area.)
3. Continuing vehicles' specific conversion problems. (There is an active research program on conversion and advisory services are being established - by equipment suppliers and on an independent basis.)

CNG's actual contribution to reducing New Zealand's dependence on imported fuel is quite modest, even if the 1985 target is achieved - which we are confident it will be.

There has been considerable debate recently about a much higher target for CNG conversions as an alternative to a gas to gasoline plant using the Mobil process - the figure generally mentioned is an additional 400,000 converted vehicles to replace the petrol that would be produced by such a plant. The Government has now approved the synthetic petrol plant. We, in any case, did not take the alternative too seriously because of the unrealistically high proportion of suitable vehicles which would have been converted and other factors.

Although modest, CNG's contribution is nevertheless a very significant one - it is something we are doing now.

To close, Figure 5 shows that with CNG, LPG and synthetic petrol, all currently aimed at replacing petrol, our next major target for greater self-sufficiency is diesel. CNG will, we hope, also prove to have a substantial contribution to make in this area.

THE ITALIAN EXPERIENCE

Gustavo Bonvecchiato
and
Pietro Magistris

SNAM, Italy

A distribution system for the sale of compressed natural gas as a vehicular fuel has operated in Italy for 40 years and now more than 250,000 dual fuel cars (CNG and gasoline) are served by 220 CNG refueling stations scattered in the North and the center of the country.

This report describes the CNG distribution network and the problems connected with station design, operation, safety, metering and billing.

The economics for station operators and car owners, as well as the pros and cons of car conversion to CNG, are also discussed.

Introduction

In Italy the use of compressed natural gas as a vehicular fuel began just before the onset of the Second World War. Its use became more popular throughout the war, managing to cope with the lack of gasoline; in fact, this was also the first way natural gas was used in this country. During those years CNG was used for public transport (railways and town buses) and private vehicles. Afterwards it was no longer used for public transport but only in private vehicles. It was shown that, as of December 31, 1980, about 270,000 to 280,000 cars were using it with around 750,000 cylinders. Now there are 220 stations selling CNG in a good part of Northern and Central Italy as an alternative to traditional fuels.

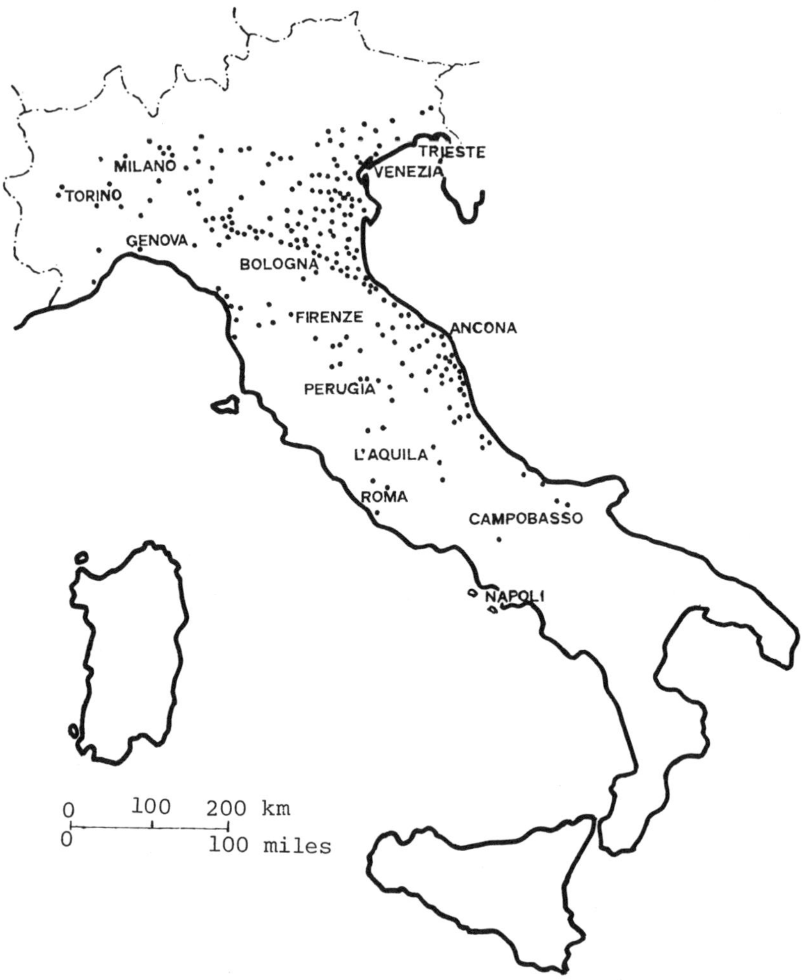

FIGURE 1. Gas Refueling Stations in Italy

Service Stations

The location of the stations throughout Italy is shown in Figure 1. SNAM supplies natural gas to 204 of them, owns 33 stations but directly operates only 14. Many small companies own or operate all the others, singularly or in groups of two or three.

There are two different types of CNG stations:

1. systems connected to the national natural gas pipeline network (compressed stations), and

2. systems not connected to it but supplied with CNG by means of cylinder trucks (transfer stations or filling stations).

Compression Stations

There are 135 compression stations, most of them built according to the flow diagram shown in Figure 2. In these stations natural gas, after having been measured with a volumetric meter, is compressed in storage cylinders by reciprocating compressors having two or three stages, according to available pressure.

Large storage cylinders, during the peak of demand, issue a quantity of CNG greater than compressor capacity and furthermore, during hours of minor sales, can refuel several cars without having to start the compressors.

The storage cylinders are divided into two groups:

- the first group remains at medium-pressure, from about 130 to 200 kg/sq cm (1800 to 2800 psi);
- in the second, on the other hand, the pressure varies from 220 to about 250 kg/sq cm (3100 to 3500 psi).

CNG flows from the storage cylinders to the refueling stalls and the quantity of stalls varies from four to ten depending on station size. For refueling vehicles a flexible line, operating on the valves situated near each stall, connects the vehicle's cylinders to the medium-pressure storage first and then to high-pressure storage. Refueling is done with a pressure of 220 kg/sq cm (3100 psi) maintained by a regulator situated at the high-pressure storage outlet. After refueling and when valves are closed, the gas left in the flexible line is discharged into the atmosphere before disconnecting the vehicle.

About eight to nine minutes are needed for all refueling operations. During the last few years, as well as the compression stations built in accordance with the diagram in Figure 2, other stations have been built according to the gas flow diagram shown in Figure 3.

In these plants, downstream from the compressors, there is only one storage area where, during operation, gas pressure varies from about 130 to about 200 kg/sq cm (1800 to 2800 psi). Refueling is done by connecting the vehicle's cylinder to the sole line coming from storage. There a monostage compressor is inserted, which completes the operation, always at a maximum pressure for 220 kg/sq cm (3100 psi), after the pressure in the vehicle's cylinders has reached that of storage, due to a simple transfer.

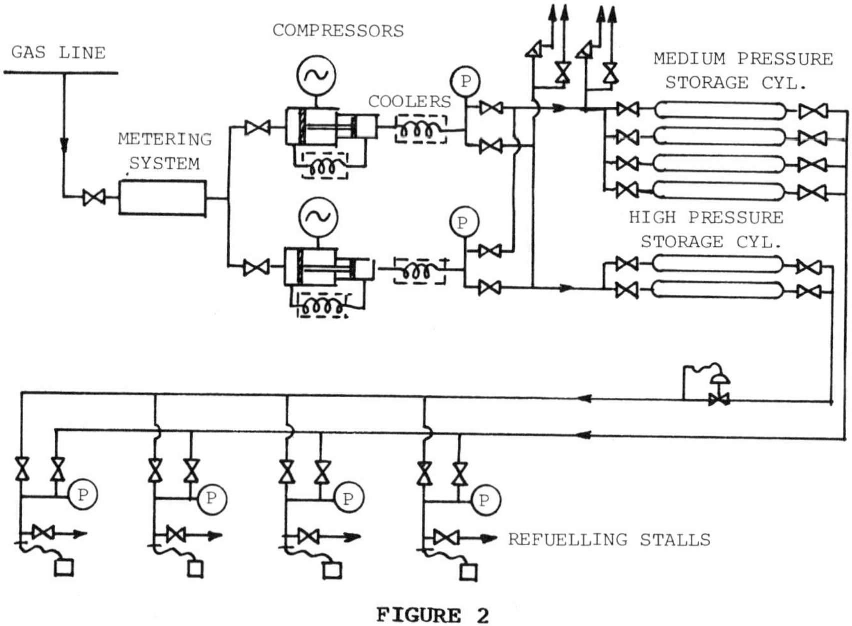

FIGURE 2

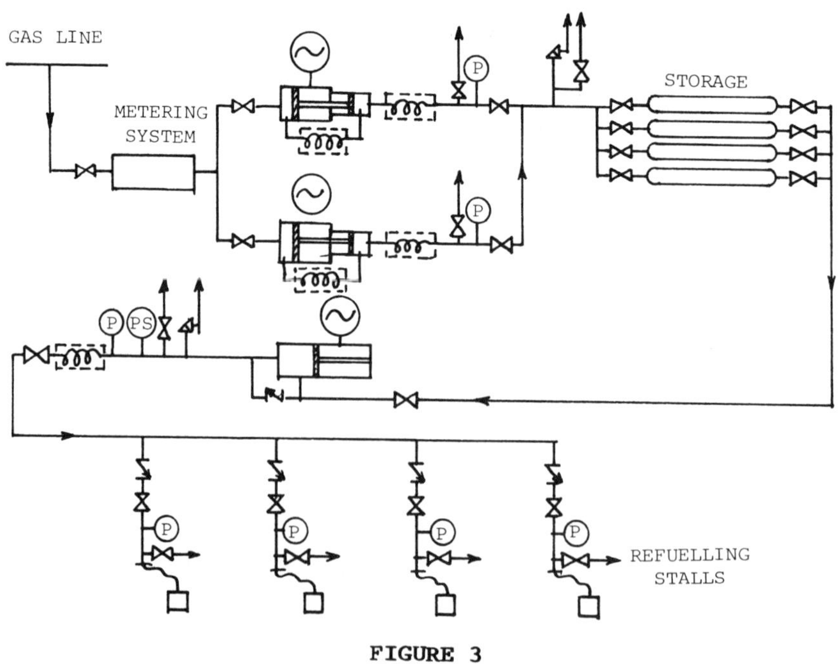

FIGURE 3

In this kind of station the regulator that ensures supply pressure at 200 kg/sq cm is not necessary, as it is obtained by controlling the monostage compressor operation by a simple pressure switch. However, in some cases a pressure regulator is installed downstream of the compressor, thus giving better control on final refueling pressure.

As can be seen when comparing the two flow diagrams of Figures 2 and 3, the second type of station is simpler, as far as the plant itself and its operation are concerned. Refueling is done by manoeuvring just one valve but it has the disadvantage of transmitting pressure pulses to the cylinders in the car during refueling, thus increasing their risk of fatigue and of being less apt to be converted to all "self service" type.

Transfer Stations

There are 85 transfer stations which are supplied with CNG by means of cylinder trucks. These are filled at compression stations equipped for that purpose as well as for sales to the public. The flow diagrams adopted in this kind of plant are different - the most common is that shown in Figure 4. As can be seen, the plants of these stations consist of:

- a cylinder truck with cylinders, divided into two groups, having a pressure of 200 kg/sq cm (2800 psi), that transports gas by road from the compression station to the transfer station and acts as a storage center at the same station during operation;
- a small storage center, permanently at the plant, where pressure varies during operation from about 130 to 200 kg/sq cm (1800 to 2800 psi);
- two monostage compressors; and
- a number of refueling stalls - normally from four to six.

One of the two compressors is at the outlet side of the permanent storage to finish refueling at 220 kg/sq cm (3100 psi) after the pressure in the vehicle's cylinders has reached, by a simple transfer, that of storage itself, and it is controlled by a pressure switch. The second compressor is used by inserting it, alternatively, between the two cylinder truck groups and then between them and the permanent storage. Thus, by using the compressor it is possible to maintain a sufficiently high pressure in the permanent storage to feed the first compressor for refueling and to empty adequately the cylinder truck.

If the operator takes care in using the second compressor, the level of cylinder truck utilization could be up to 80% of its total

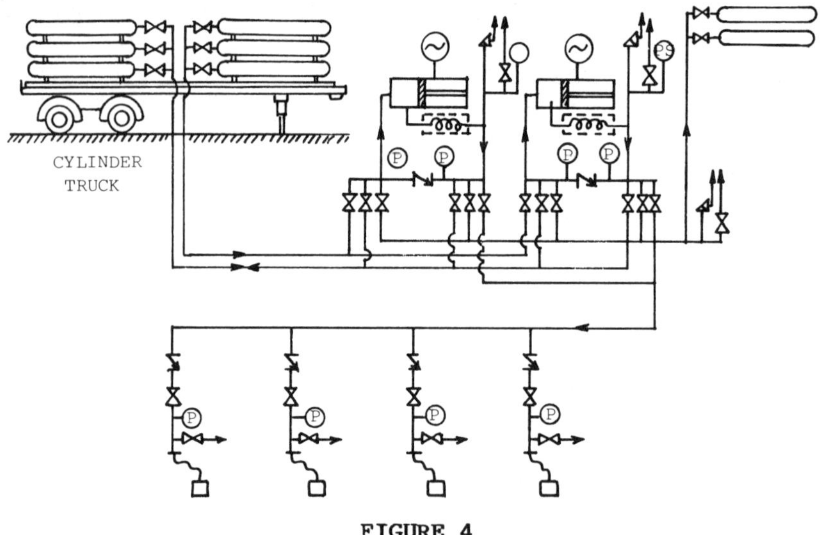

FIGURE 4

transport capacity. The two compressors are interchangeable in the previously described functions and can also be used in parallel for particular operating requirements.

Station Supplied With LNG

To complete the description of the various kinds of CNG stations set up in Italy, one should remember that for about two years SNAM successfully operated a station supplied with liquefied natural gas (LNG). This plant was implemented according to the flow diagram shown in Figure 5 and consisted of:

- a 30 cu. m. cryogenic tank for LNG (190 bbl);
- a pump that compressed LNG up to 250 kg/sq cm (3500 psi);
- a system of two air evaporators and electrical heating in series; and
- a group of four refueling stalls.

This plant was periodically refueled with a 20 cu.m. (125 bbl) cryogenic fuel-truck and CNG was obtained by means of compression at 250 kg/sq cm of LNG that was later brought to the gaseous state in the evaporators and accumulated in the group of large cylinders ready for sale. To refuel vehicles the sole pipe coming from storage was used; in it a regulator was inserted to ensure an output pressure of 220 kg/sq cm (3100 psi). This plant was set up as an experiment, was dismantled due to difficulties that arose in LNG supplies and was transformed into a transfer station.

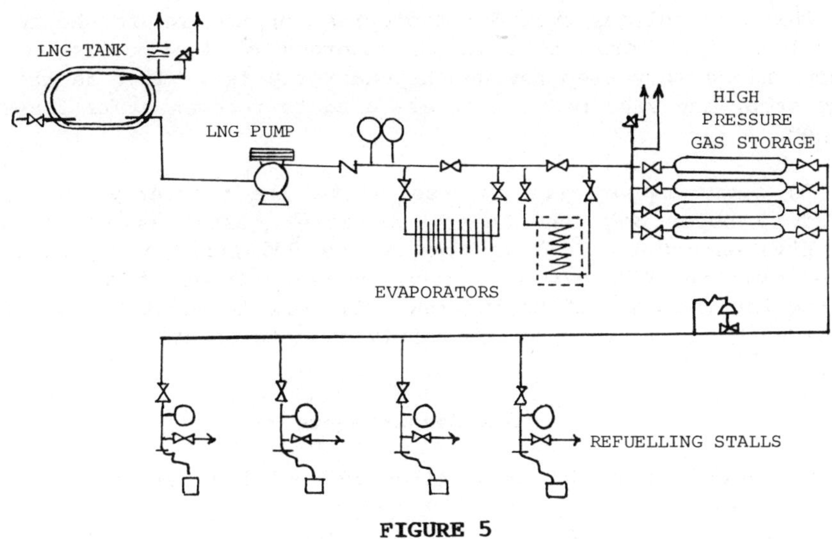

FIGURE 5

Laws, Regulations and Safety Guidelines

The CNG stations are built in accordance with the Home Office (Ministero degli Interni) regulations issued in 1954 and 1959 and kept under control by the Fire Brigade (Vigili del Fuoco). These regulations are particularly strict, as they call for the use of reinforced concrete structures for civil works, refueling stalls, compressor premises, storage cylinder premises, and stalls for the cylinder trucks. These regulations also impose a safety distance of up to 50 meters (165 ft) from any building close to the plant, and of up to 15 meters (50 ft) between plant elements as, for example, between compressor premises and those used for storage.

As a consequence, the service stations selling CNG are installed in the outskirts of the towns or along main roads where the required area could be found easily and cheaply. Because of the expansion of the towns during the last decades some stations built in congested areas had to be relocated out of town. This explains how in some new buildings very old compressors have been installed.

The possession of CNG cylinders to be used as fuel tanks in cars is regulated in Italy by a law which obliges all users to pay, every three months, a small sum for each cylinder. This revenue feeds a special fund, managed by the Agency "Fondo Bombole Metano" which, without other payments or charges, assures to cylinder users free periodical testing of cylinders (and their substitution if discarded) and insurance for damages for accidents arising from cylinder use.

This law obliges cylinder factories, or the owners who re-sell their used cylinders, to give the address of the buyers to this Agency unless they keep paying the quarterly fee. In case the payments stop, the same means that are used to recover state taxes are applied.

This may appear very oppressive but the quarterly fee is adjusted to cover only the Agency operative costs, as it is not a lucrative enterprise. Today the fee is 990 lire per cylinder and it relieves the CNG car user from the temptation of avoiding the testing for the fear of buying new cylinders in substitution of the discarded ones. Its impact on safety is noteworthy.

Gas Measurements

The metering of CNG is obtained by applying the formula:

$$V = \frac{C}{1000} \times (220 - Pr - Ps) \times 1.25$$

where:

- V is the quantity of gas delivered during refueling, expressed in standard cu meters (at atmospheric pressure and 15° C);
- C is the capacity of the cylinders installed on a buyer's car and is expressed in liters;
- 1.25 is the value normally used as coefficient of natural gas compressibility;
- 220 is the value of pressure upon completion of refueling, expressed in kg/sq cm;
- Pr is the residual pressure, that is the pressure of the gas eventually contained in the cylinders before refueling (its value is rounded off to the nearest 10 kg/sq cm);
- Ps is the increase given to residual pressure to compensate empirically the influence of the increase in temperature of gas during refueling; when using the formula, this increase assumes the following values:

Value of Ps	Pr From....	....To (kg/sq cm)
20	0	30
15	40	70
10	80	130
5	140	150
0	above 150	

For the practical application of this formula there are tables that facilitate calculation of the volume of CNG delivered as a function of the "C" and "Pr" values, shown at the beginning of refueling. However, nowadays a small computer is used in several stations which, when fed with the "C" and "Pr" values, gives the sum to be paid. At the same time sales are systematically recorded and automatically the daily plant accounts are obtained.

Economics

Fuel Sales and Prices

CNG sales in Italy in the year 1980 were equal to 312 million cu. m. (11 billion cu ft), while sales of gasoline were equal to 16.5 billion liters (4.35 billion gallons). If one assumes that 1 cu. m. of methane equals 1.1 liters of gasoline, CNG consumption represented only 2.1% of gasoline consumption.

Figure 6 shows a 20 year comparison, in real terms, between price differences of gasoline and CNG and the volumes of CNG sold for vehicles. Figures 7 and 8 show the selling prices of gasoline, diesel fuel, LPG (liquefied petroleum gas for vehicles) and CNG during the last ten years; in one there are the absolute values, in the following figure the prices in constant terms. In both diagrams prices refer to net calorific values (NCV) in order to have better comparisons. From these figures we can see that the last decade can be split into three periods. In the first one, from 1971 to 1973, although the sale price of CNG allowed a two-thirds cost reduction compared to gasoline, there was only a slight sale increase for CNG because the competition of the two other

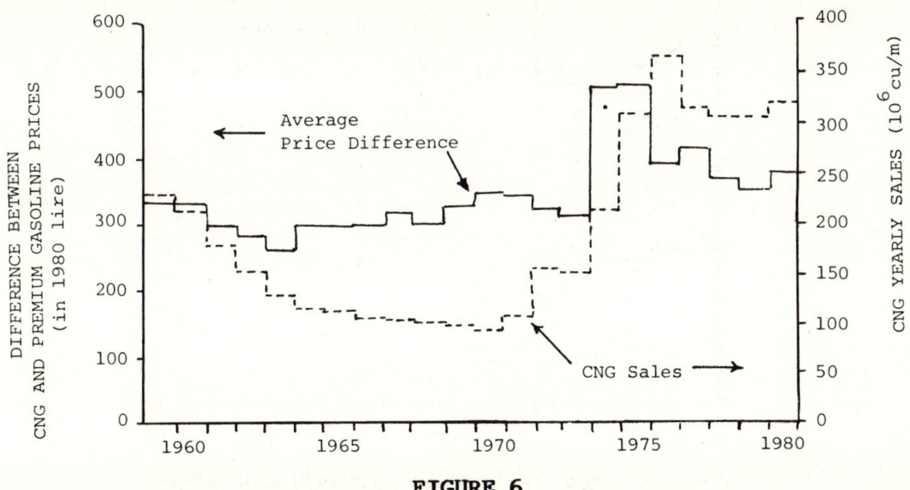

FIGURE 6

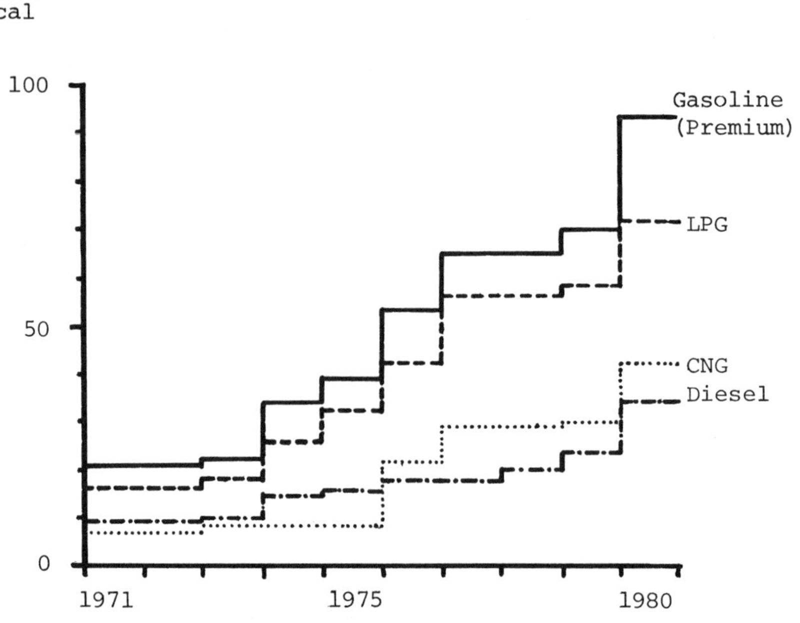

FIGURE 7. Yearly Average Prices (1980 Lire)

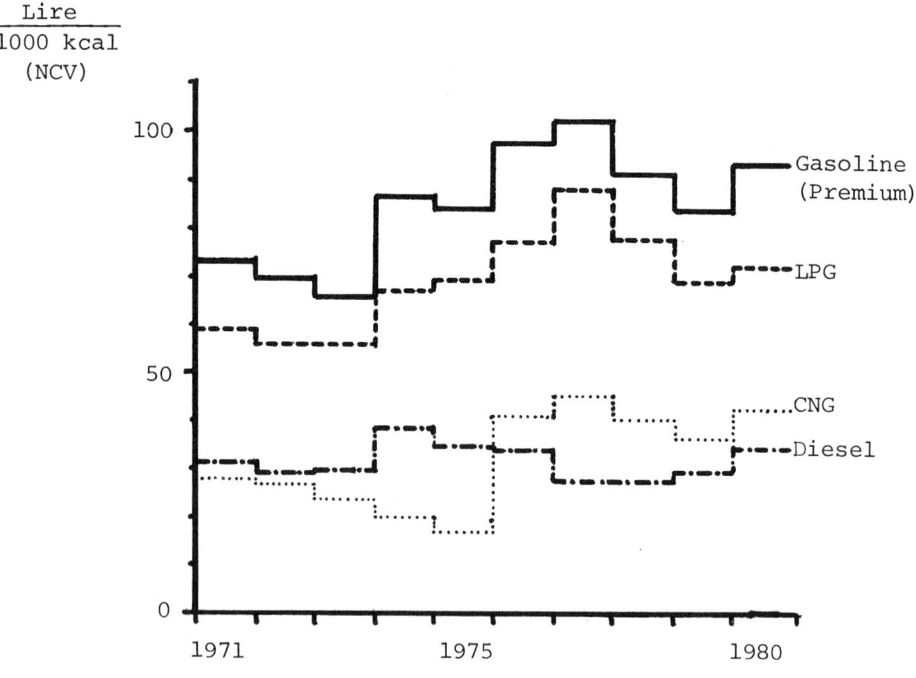

FIGURE 8. Yearly Average Values (1980 Lire)

alternative fuels was strong. By contrast, in the second period, from 1973 to 1976, a strong increase in CNG sales is shown, as a consequence of the very steep increases in the price of all other fuels, which assured the CNG users of a saving of 80% compared with gasoline use, 70% with LPG and 50% with diesel fuel. In the last period CNG sales stabilized at around 300 million cu meters per year (10.5 billion cu ft), although the saving was still high, because in this period diesel fuel was the cheapest notwithstanding the heavy increase in the yearly road tax imposed on diesel cars. This is equivalent to a price increase but it is not included in the user prices shown here as it cannot be referred to the sale price.

For a better understanding of the Italian market situation, here are the user prices in different units at the end (December 31,1980) of the examined period:

Fuel	Retail Sale price Lire/Liter	Equivalence in US $ (at Dec '80 exchange rate)		Total Fiscal charges, as percentage of sale price
		$/MMBtu	$/US gallon	
Gasoline (premium grade)	850	30.37	3.42	64.7
Diesel Fuel	337	10.61	1.36	19.7
LPG (vehicular use)	555	23.94	2.24	62.3
	Lire/cu m		$/1000 cuft	
CNG	425	13.81	12.81	45.3

Recent price increases are due to crude cost increases, thus reducing the percentage of fiscal charges. The increase of the exchange rate of the US dollar now induces an apparent reduction of retail prices.

Service Station Investments

A CNG refueling station with the following characteristics:

- Yearly sales: 3 million cu.m.(10.5 million cu.ft/y);
- 8 refueling stalls;
- Gas pressure at station inlet: 7 kg/sq cm (100 psi);
- Max outlet pressure: 250 kg/sq cm (3500 psi);
- 3- three stage electric compressors - 2 operational and 1 spare (500 cu m/h each = 300 cuf/min);
- Storage pit with 7 cylinders of 1 cu m capacity each;
- Electric substation with a 500 kWA transformer;

- Metering cabin including regulating system and gas preheating,

requires the following investments (Summer 1981 Costs in million lire - average exchange rate: 1 US dollar = 1200 lire, interest during construction and TVA not included):

- civil works, excluding the cost of acquiring land = 150
- mechanical machinery and related works = 350
- electrical components = 70.

In the above costs are included land preparation, office buildings, fencing, fire system, water connection. These costs refer to a CNG station built according to the very strict Italian regulations.

Service Station Operating Costs

The operation of CNG stations is considerably more expensive than that of the traditional gasoline refueling station because the plant is more complex. It requires more maintenance and there are consumptions of electrical energy, oil, and water for the compressors.

Compression

In relation to the pressure available in the compression station, the average specific consumption of electric energy is the following:

Pressure	(kg/sq cm)	5	10	20	40
	(psi)	70	140	280	570
Consumption	(kWh/cu mt)	0.28	0.18	0.15	0.12
	(kWh/1,000 cu ft)	8	5	4	3.5

In transfer stations, which operate on already compressed gas, there is a specific energy consumption of about 0.04 kWh/cu m (1.1 kWh/1000 cu ft).

Labor Charges

These are higher because of:

a) the considerable length of time needed for refueling;
b) the small quantity delivered at each refueling operation (on an average from 15 to 20 cu. m; 500 to 700 cu ft) and;
c) the necessity, due to lack of handy CNG measuring devices, to separate the refueling operation from the payment operation; this calls for one or more operators at the refueling stalls as well as someone at the cash desk.

In the station described above, which operates 16 hours per day and 365 days per year, 10 people plus a head of the station are required for the operation. Maintenance is not the task of these station operators.

Water Consumption

In a moderate climate (North Italy) for cooling the compressors in closed cycle and for general purposes (no car washing is provided in the refueling station), water consumption varies according to the season, but the yearly average is from 2.5 to 3 cubic meters per day or per 10,000 cu m of gas sold.

Other

Lubricating oil for the compressor is required and maintenance, labor and spare parts depend on station age and type. Usual figures for conventional machinery can be applied.

Car Owners' Costs and Troubles

CNG is used to fill vehicles owned by private persons who for different reasons, travel more than the average and who choose this fuel to reduce the fuel cost per mile. It is, however, in any moment possible to change to gasoline as the gasoline tank is always left in the converted cars. CNG use also reduces the wear of the engine and mileage between oil changes can be safely doubled.

These economical advantages are indispensable to compensate not only for the cost of transforming the vehicle but also for some disadvantages that the user must accept when using CNG. These disadvantages, of which the Italian users complain are:

- Limited range (equal to 50 to 60% of range when using gasoline);
- Engine power loss (more like a reduction in acceleration than in maximum speed);
- Size of cylinders (thus reducing trunk space);
- Weight of the cylinders (equal to 8 to 10% of vehicle weight) in some cases reduces in accordance with the road code the number of allowed passengers;
- The imperfect interchangeability between CNG and gasoline (due to the variation caused by the insertion of the CNG mixer on the carburetor and by the change in the anticipated ignition time of the spark plugs);
- Limited number of refueling stations of the distribution network; and
- time required for refueling (8 to 9 minutes).

In Italy the cost of a conversion to CNG ranges (summer 1981) from 700,000 to 900,000 lire, according to the type of car and number and type of installed cylinders (the lightweight ones are more expensive). In Italy it is an easy job usually done by very small, specialized enterprises. Required components are easily and readily found in the Italian market, usually as kits.

Conclusions

On the basis of the Italian experience we can state that, for the use of CNG as an alternate to traditional fuels, all the technical and operative problems relevant to its distribution and use have been solved in a satisfactory way and all the required material is available in the market. Feasibility studies for different local conditions, adjusting the investments and the operating costs given in this paper, can be made from the viewpoint of CNG distributors, of car users and of a country as a whole. The most important parameter is, however, the amount of the fiscal charges imposed on automotive fuels. They can be suddenly increased or abolished, altering the price difference between CNG and gasoline and upsetting the results of the feasibility studies. This means that a simple fiscal law can introduce or prevent the use of CNG as an automotive fuel in a country. Another important factor is the impact of state bureaucracy: laws, regulations and mandatory standards, new or existing in other fields, if they affect CNG use, can also influence the costs and thus the results of the feasibility studies.

APPENDIX I

The following pictures represent some typical aspects or components of a few of the 220 stations in operation in Italy.

Figures 9, 10, 11 and 12:	show the general view and the stalls of stations located respectively in Bergamo, Bologna, Arezzo, Florence.
Figure 13:	stalls and cars during refueling (Florence).
Figures 14 and 15:	cylinder trucks on their way to supply the transfer station of Traversetolo.
Figure 16:	a cylinder truck supplies the cars on the background in the small transfer station of Casalmaggiore.
Figures 17, 18 and 19:	reciprocating compressors, driven by electric motors or by gas engines, in the factory and in Parma station.
Figure 20:	compressor room of the Florence station.
Figure 21:	filters, reducers and meters in the metering cabin of the Bergamo station.
Figure 22:	pit of storage cylinders used as water tank at Bologna station.

FIGURE 9

FIGURE 10

FIGURE 11

FIGURE 12

FIGURE 13

FIGURE 14

FIGURE 15

FIGURE 16

FIGURE 18

FIGURE 17

FIGURE 19

FIGURE 20

FIGURE 21

FIGURE 22

THE JAPANESE EXPERIENCE

Munenobu Tanaka

Chairman of the Department of Mechanical Engineering
University of Tokyo

In this chapter I would like to explain the position of methane as a primary energy source in Japan, with regard to its share, price and use. Next I will explain the past Japanese experience with methane for transportation use, although this experience is quite old and out of date.

Let me begin with the first problem. To make the methane role clear, it is important to understand the total energy picture in Japan. The figures in Table 1 show the petroleum equivalent in megatonnes and from this we can see that imported energy is increasing every year. Notice also that more than 70% of total energy is derived from petroleum, almost all of which Japan imports from abroad.

According to the 1976 statistics, 17% of the total energy was consumed for transportation uses, including 3% for international transportation. With respect to petroleum, 29% was spent for transportation, most of which was imported from abroad; so Japan as a nation is importing almost all of its energy from abroad, as much as 86%. An alternative fuel for transportation is very important.

Next, I would like to explain the methane situation in Japan. Table 2 shows the supply and the demand for natural gas and its availability in Japan. We can see that imported LNG is increasing and used mostly for power stations and city gases. Table 3 shows the countries from which Japan imports most of its LNG.

Next, let us examine the supplies of energy in Japan. Table 4 shows the price of LNG increasing with the price of crude oil, coal and LPG.

TABLE 1. PRIMARY ENERGY IN JAPAN IN MEGATONNES PETROLEUM EQUIVALENT

	1975	1976	1977	1978	1979
Total Primary Energy Supply	366(100)	384(100)	384(100)	385(100)	409(100)
Natural Gas (domestic)	3(0.8)	3(0.8)	3(0.8)	3(0.8)	3(0.7)
LNG (import)	7(1.9)	8(2.1)	11(2.9)	16(4.2)	20(4.9)
Water Power	21(5.7)	22(5.7)	19(4.9)	18(4.7)	21(5.1)
Nuclear Power	6(1.6)	8(2.1)	8(2.1)	15(3.9)	17(4.2)
Coal	60(16.3)	59(15.4)	57(14.8)	53(13.8)	57(13.9)
Petroleum	268(73.2)	284(74.0)	286(74.5)	281(73.0)	291(71.1)
etc.	1(0.3)	--(--)	--(--)	--(--)	--(--)

Table 5 illustrates the price per heat value. From this we can see that LNG is almost the same price as petroleum. Therefore, LNG is not a less expensive energy source in our country. This fact may be a major obstacle to promoting methane-powered taxis in Japan.

Table 6 shows the future energy plan of the Japanese government, with its supply target for various alternative fuels in 1990. We expect to get 20% of our total energy from LNG. The right side shows the status in 1978 for reference.

Tables 7 and 8 show the energy and transportation energy outlook projected to 1990: Table 9 indicates that whereas the shortage of diesel fuel will become important in Japan, the shortage of gasoline will not be so serious.

Now I should like to offer some data on the Japanese past experience with methane buses. During the last world war, Japan developed many cars running on alternative fuel. For example, in 1938 both a charcoal car and a kerosene car were developed, in 1939

TABLE 2. SUPPLY AND DEMAND OF NATURAL GAS IN JAPAN

				1975	1976	1977	1978	1979
Natural Gas ($10^6 m^3$)	Coal Mine Gas	Supply	Production	332	318	270	288	279
		Demand	For City Gas	14	13	13	14	13
			For Mining	6	8	7	7	7
			For Energy Industry	213	232	198	203	205
			Etc.	2	2	2	2	2
	Oil Field Gas	Supply	Production	2446	2604	2775	2583	2351
		Demand	For Electric Power	237	394	664	646	614
			For City Gas	624	661	662	646	611
			For Mining	373	343	295	198	139
			For Energy Industry	27	26	24	24	23
			Etc.	1	2	6	18	20
			Not for Energy	1110	1105	1004	963	871
LNG (10^3 ton)		Supply	Import	5059	5977	8386	11719	14858
		Demand	For Electric Power	3326	3920	5703	8603	11708
			For City Gas	1614	1955	2409	2692	3053
			For Mining	--	17	92	213	191
			For Energy Industry	7	17	20	11	17

TABLE 3. JAPAN LNG IMPORT
(1000 t)

	1975	1977	1978	1979
TOTAL IMPORT	5,038 (100%)	8,391 (100%)	11,719 (100%)	14,569 (100%)
BRUNEI	1,671 (33%)	5,295 (63%)	5,329 (45%)	5,567 (38%)
AMERICA	3,367 (67%)	1,013 (12%)	958 (8%)	986 (7%)
INDONESIA	0	1,377 (16%)	4,246 (36%)	6,546 (45%)
U.ARAB EMIRATES	0	706 (8%)	1,185 (10%)	1,462 (10%)

TABLE 4. COST OF IMPORTED LNG (cost per weight)

	Cost of Imported L.N.G. (yen/ton)					Crude Petroleum (yen/kl)	Coal (yen/ton)	L.P.G. (yen/ton)
	Average	BRUNEI	AMERICA	UAE	INDONESIA			
1979 Jan.	24340	23691	22032	22210	26932	17049	------	28697
1980 Jan.	47953	38818	42060	39077	57984	43721	10438	67177
1981 Jan.	61951	61927	62416	64038	61337	45904	12375	68549

TABLE 5. FUEL COST PER HEAT VALUE (yen/1000 kcal)

	L.N.G.	Crude Petroleum	Coal	L.P.G.
1979 Jan.	1.83	1.81	-----	2.39
1980 Jan.	3.61	4.65	1.68	5.60
1981 Jan.	4.66	4.88	2.00	5.71

<u>Calories for Various Fuels</u>

LNG .... 13300 kcal/kg
Coal 6200 kcal/1kg
Crude Petroleum .. 9400 kcal/kg
LPG .... 12000 kcal/kg

TABLE 6. SUPPLY TARGET FOR VARIOUS ALTERNATIVE FUELS

Alternative Fuels	1990 x 10^4kl Petroleum		Ref. 1978 x 10^4kl Petroleum	
Coal	12300	35.4%	5681	50.9%
Nuclear Power	7590	21.8%	1542	13.8%
Natural Gas	7110	20.4%	1940	17.4%
Water Power	3190	9.2%	1941	17.4%
Terrestrial Heat	730	2.1%	16	0.2%
Other Alternative Energy	3850	11.1%	38	0.3%
Total	35 x 10^8kl	100.0%		100.0%

there was an acetylene car and, in 1941, a natural gas car. Of these alternatives, the charcoal-fueled cars were the most-used. However, the natural gas cars proved the most successful, and have the greatest survival rate. In 1941, for the first time, 70 natural gas buses were used for commuting in Nagata Prefecture. The number of buses using natural gas there increased each year until 1961, when a maximum of as many as 500 buses were used. However, they disappeared with the advent of larger buses.

A certain high-ranking official of the Ministry of Transportation in Japan told me recently that there is now no methane car in Japan. I have to say, therefore, that our experience with methane cars was quite a few years ago and, from the technical point of view, the technology was primitive. But I do have some interesting material from the Nagata Bus Company.

Figure 1 shows the natural gas vehicle system and Figure 2 illustrates the compressor system for a natural gas bus.

I was deeply impressed by the enthusiasm, the speeches and the display at the 1981 international conference in Vancouver and I think something should be done to restart this movement in Japan.

TABLE 7. ENERGY OUTLOOK IN 1990

			Quantity of Products and/or Imports	(%)	I.A.R.[(1)]	Energy Supply	Energy (%)	Primary Energy		Secondary Energy		
									Coal		Electric Energy	Pet. Products LPG
SUPPLY	Water Power	$5.3x10^7$ kW	3190	4.6	0	3190	4.6			3190	3190	
	Terrestrial Heat	$3.5x10^6$ kW	730	1.0	0	730	1.0			730	640	
	Natural Gas	$7.6x10^6$ kW	760	1.1	0	760	1.1	600		160	79	
	Coal	$1.6x10^8$ ton	12300	17.6	0	12300	17.6	1463	1463	10837	3606	
	Nuclear Power	5.1–5.3 $x10^7$ kW	7590	10.9	0	7590	10.9			7590	7590	
	L.N.G.	$4.5x10^7$ ton	6350	9.0	0	6350	9.0	85		6265	4575	
	New Energy	$3.9x10^7$ kl	3850	5.5	0	3850	5.5			3850	359	2809
	Oil & L.P.G.	$3.5x10^8$ kl	35190	50.3	0	35190	50.3	130		35060	6402	27949
	Total Energy		69960	100	0	69960	100	2278	1463	67682	26441	30758
	Supply Loss		-3226									
	Total Demand		66734				100	2278	1463	64456	25150	29404
DEMAND	Industry		41440 35842				62.1 53.7	2228 1905	1423 1390	39212 33937	15821 12451	15828 14108
	Transportation		9910				14.9	0	0	9910	489	9420
	For Civil Life		15384				23.0	50	40	15334	8840	9156

(1) Inventory Adjustment Rate

TABLE 8. TRANSPORTATION ENERGY OUTLOOK FOR 1990

ENERGY	CONSUMPTION (A)	SUPPLY (B)	SHORTAGE (C)=(A)-(B)	SHORTAGE RATE (C)/(A) (%)
Gasoline $\times 10^3$kl	40582	36384	4198	10.3
Jet Fuel Oil $\times 10^3$kl	3301	2337	964	29.2
Diesel Oil $\times 10^3$kl	24530	17743	6787	27.7
Fuel Oil A $\times 10^3$kl	3243	2501	742	22.9
Fuel Oil B.C. $\times 10^3$kl	8742	8742	0	---
LPG $\times 10^3$kl	2894	2894	0	---
Electric Power $\times 10^6$kwh	13568	13568	0	---

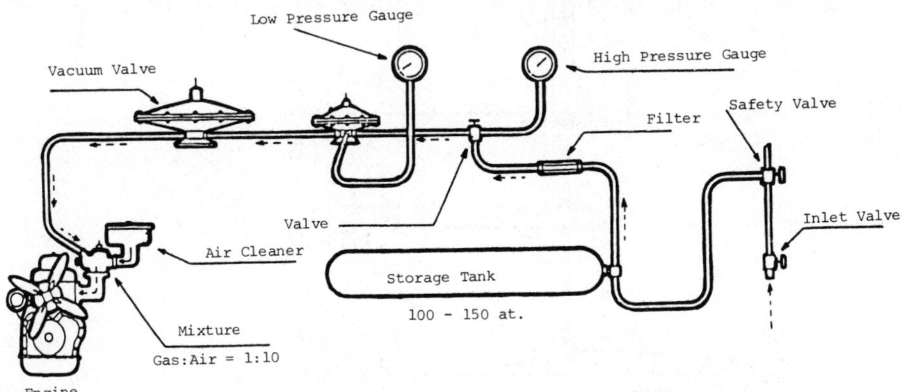

FIGURE 1. Natural Gas Vehicle Engine System

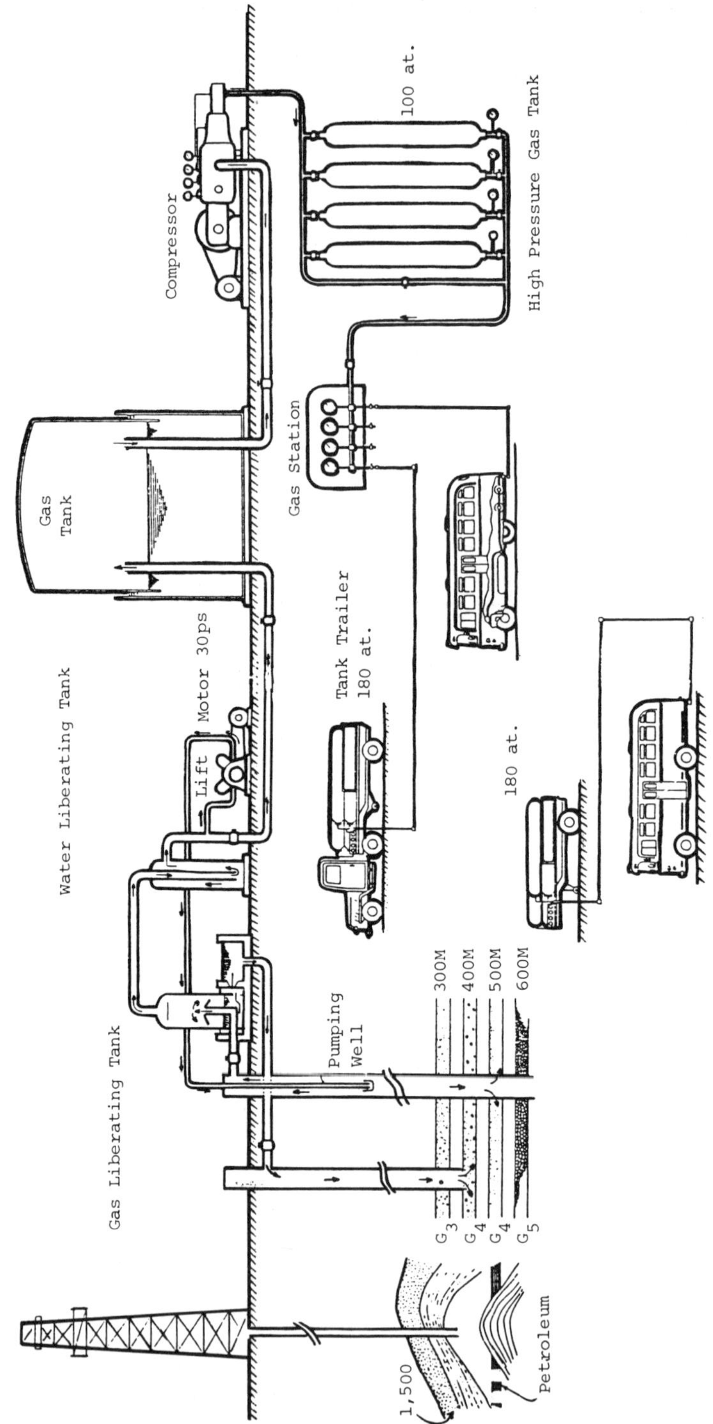

FIGURE 2. Natural Gas Vehicle Engine System

THE NETHERLANDS EXPERIENCE

Jouke van der Weide

Research Institute for Road Vehicles TNO

Delft, The Netherlands

Introduction

This is a short survey of the activities of the TNO (Research Institute for Road Vehicles) organization in the Netherlands in the application of methane in diesel and spark ignition engines. This survey will highlight the importance of alternate fuels in the future, the various applications of LPG in cars and heavy duty vehicles as well as the future of LPG and natural gas.

The various application possibilities of natural gas in spark ignition engines of cars and vans will be mentioned; then the application in heavy vehicles will be discussed. Finally, the safety with respect to application in vehicles will be discussed.

Background Information

Firstly, here is some background information with regard to the need for alternative fuels. The American Civil Aircraft Industry has done a study about the future price development of jet fuel A, which shows that in 1980 the price was $0.83 (US) and in 1990 it will be $2.80 (US). (These figures have been corrected for inflation.) The expectation is that for the jet fuel the price will be 3 1/2 times higher within ten years. It must be realized that what applies to jet-fuel also applies to diesel fuel - they are rather similar fuels - and, in fact, applies to more or less all liquid fuels. Thus liquid fuels will become expensive in the future.

Figure 1 illustrates the future expectations of the differences between demand and production of liquid fuels. Some shortages are expected in 1982 through 1984 and from 1985 the demand for liquid fuels will exceed the production indefinitely. The result of such a difference between demand and production is known from the Rotterdam spot market. When this situation occurred two years ago the spot market price doubled, in comparison with OPEC prices, within a few months.

Future Availability of LPG and Natural Gas

A very strange situation exists in the oil fields: a lot of gaseous fuels are wasted by flaring off. This is so-called associated gas, which is gas dissolved in oil as long as the oil is under high pressure in the soil. By releasing this pressure the gaseous components boil out in the same way carbon dioxide does when a bottle of soda water is opened.

Satellite photos of the Middle East at night have been published showing the enormous amounts of fuel flared off. (Figure 2) In Saudi Arabia alone it is estimated that about 100 million tons of gaseous fuels per annum are flared off. Theoretically the amount of flared off gas in the Middle East area represents about two-thirds of the required energy for transportation in Western Europe.

LPG Application in Vehicles

LPG is widely used nowadays in the Netherlands as an automotive fuel and the number of LPG vehicles, estimated in 1981 at approximately 450,000, is still increasing. These LPG vehicles can

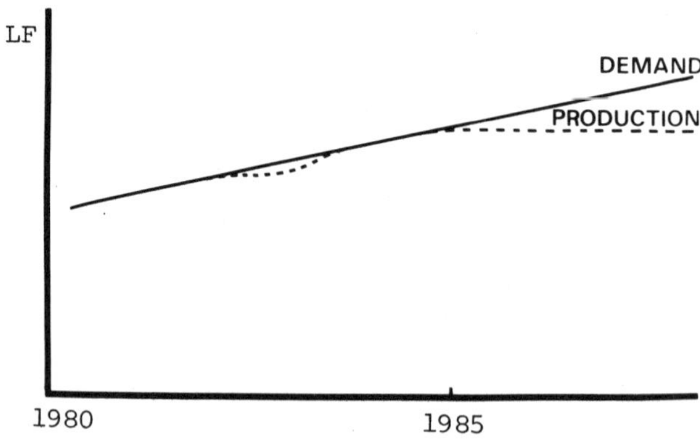

FIGURE 1. Differences in demand and production of liquid fuels in the future.

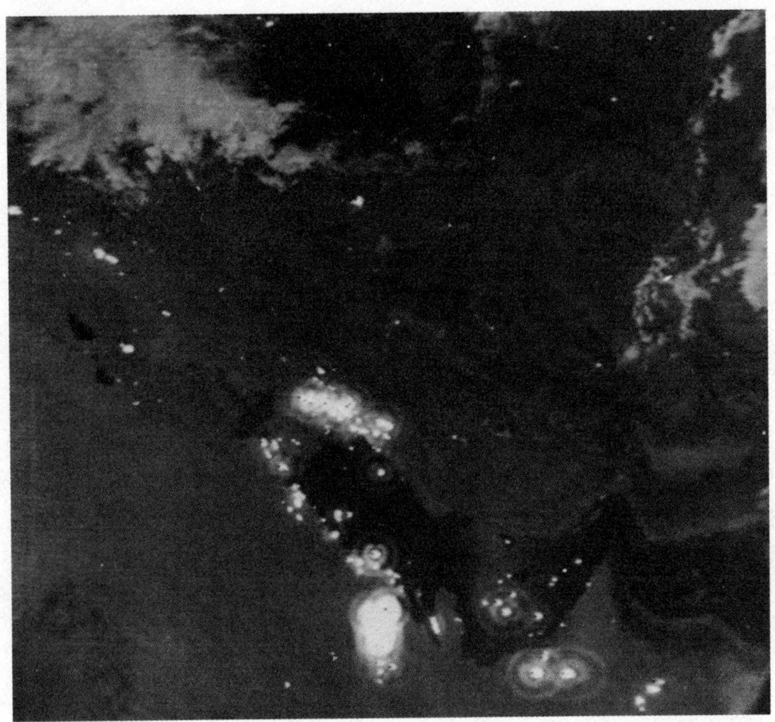

FIGURE 2. **Satellite photo of the Persian Gulf area at night. Top, Caspian Sea. Middle, below, Persian Gulf. Note the large sizes of the flares of associated gas in comparison with a city like Tehran (white spot just beneath the Caspian Sea).**

be refueled at approximately 2500 public filling stations. These filling stations all combine the sale of LPG with conventional fuels such as gasoline and diesel oil. In fact, the use of LPG is extending to heavy duty vehicles originally equipped with diesel engines. Figure 4 shows a prototype city bus in Amsterdam running on 100% LPG. Vienna also has a long history with regard to the application of LPG in buses: starting in 1963 buses there used mixed diesel-LPG systems and, in 1976, some of these changed to run on 100% LPG.

Nowadays half of the city bus fleet (approximately 200 buses) is still running on the mixed diesel-LPG system while the other half runs on 100% LPG. Most of this latter group are large double decker buses.

Apart from the economic incentive, a particular advantage of the application of LPG in these vehicles is the lower exhaust

FIGURE 3. At a Dutch filling station, all dispensing units look similar.

FIGURE 4. Amsterdam city bus running on 100% LPG.

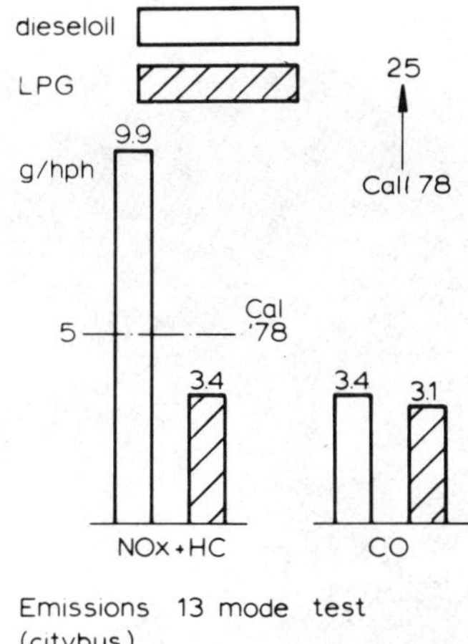

FIGURE 5. 13-Mode emission data from 100% LPG bus.

emission. Not only is the exhaust free of black smoke, but the level of nitrogen oxide is substantially lower. (Figure 5)

With the mixed diesel-LPG system, the diesel injection pump is re-adjusted to approximately 70% of the nominal delivery. The remaining 30% of the diesel fuel is substituted by LPG carbureted into the air intake manifold. The combustion of the fuel air mixture in the engine is still introduced by the injection of diesel fuel. Therefore, in a diesel engine the power and torque characteristics will be similar to the original engine. Figure 6 shows the hardware of such a system mounted on an engine. On top of the intake manifold is placed a venturi mixing the LPG with air. The aluminum part in front of the injection pump is the LPG control valve.

Natural Gas Application in Vehicles

Natural gas can be stored in a vehicle in three ways:

1. Compressed natural gas (CNG) is usually stored in steel cylinders under a pressure of approximately 200 bars.

2. Liquefied natural gas (LNG) is stored in cryogenic vessels under a relatively low pressure at a temperature of -164°C. Figure 7 shows such a cryogenic LNG tank fitted in the trunk of a car. The conversion was done by TNO and the experience showed that it operated quite well. However, although the range was sufficient the hardware required for the cryogenic systems was then very

FIGURE 6. Mixed diesel-LPG system. Note venturi in the intake and extra component on front at the injection pump.

FIGURE 7. LNG tank in the boot of a car.

expensive and did not allow an economic application.

3. Natural gas can be converted into methanol by adding one oxygen atom to a methane molecule. Although this is more complex than it sounds, it is done in the Netherlands on a large scale.

In general the exhaust emissions of gaseous fueled cars are much cleaner than gasoline exhaust emissions. TNO conducted an extensive test program with seven European cars fueled respectively with leaded gasoline, unleaded gasoline, LPG and natural gas. Figure 8 shows the average exhaust emission measured according to the ECE 15 cycle for the different fuels. Notice the substantial reduction in carbon monoxide emission. However, it must be noted that the program was executed about six years ago and that the gasoline emissions are not representative of today's cars.

Natural gas can also be used in diesel engines, generally used in trucks and buses. In principle, there are three possible ways to do this.

1. 100% Diesel Replacement by CNG

In this case, the diesel engine has been converted into an Otto engine which means the compression ratio has to be lowered, and spark ignition and NG carburetion systems have to be added.

2. 85% Diesel Replacement by CNG

In this system, approximately 85% of the diesel fuel is sub-

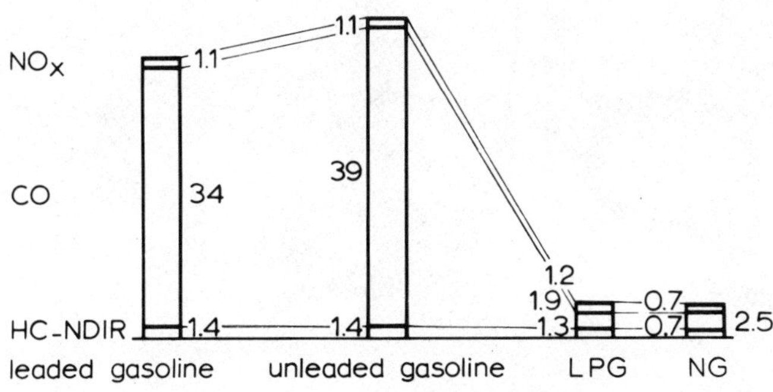

FIGURE 8. Exhaust emissions averaged from seven European cars in ECE 15 cycle, fueled with different petrols, LPG and natural gas.

stituted by means of carbureting natural gas into the air intake manifold. The remaining 15% of the diesel fuel is used to introduce the combustion in the engine.

3. 35% Diesel Replacement by CNG

This third method substitutes approximately 35% of the diesel fuel by carbureting natural gas into the air intake manifold. The remaining 65% of diesel fuel is delivered to the engine in the normal way.

Figure 10 shows a typical LPG tank arrangement under a city bus. These contain 320 liters of LPG. In heavy trucks or buses the substitution of diesel oil with CNG will create problems. Even if the engine had the same energy efficiency the volume needed for the storage of CNG is approximately 13 times the volume needed for the storage of diesel fuel. As well, using steel cylinders for the storage of CNG increases the weight of fuel and tanks by approximately 6 times. In this case the vehicle would have the same range as on the original fuel. The extra weight will cause a loss in actual pay load for the vehicle and, in the case of buses, a loss of some passenger places. Replacement of the steel cylinders by composite pressure vessels would reduce the extra weight approximately 3-fold. If a loss in payload or passenger places is not acceptable the amount of fuel has to be reduced. This will cause a reduction in range of about one-third.

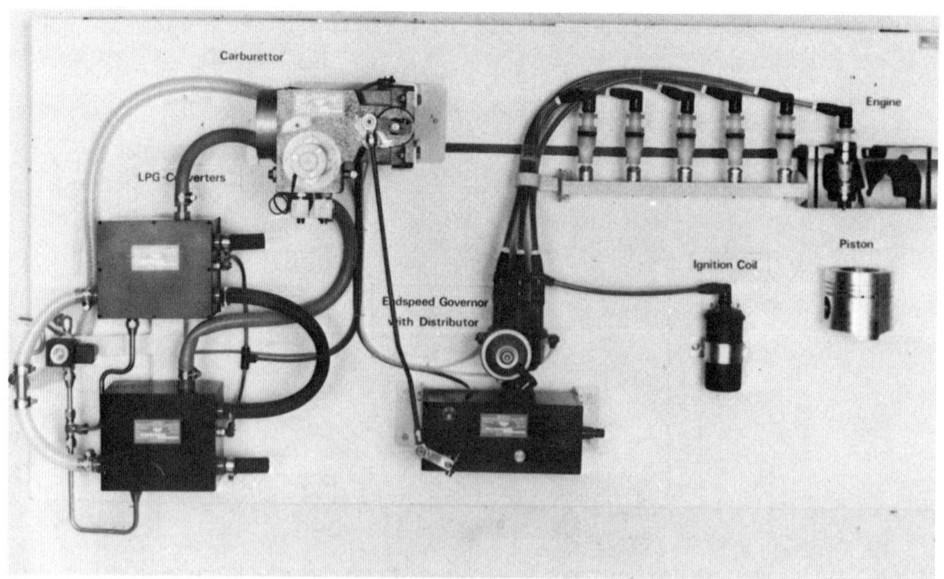

FIGURE 9. Hardware components for a 100% gaseous system on diesel engines (conversion to spark ignition).

FIGURE 10. LPG tank arrangement under a city bus.

For these reasons it is often said that the application of the so-called mixture operations is more feasible. With mixture operation about one-third of the diesel fuel is substituted by a gaseous fuel, LPG or CNG. In this case the extra volume required and the extra weight will cause a much smaller loss in payload or passenger places.

The control of the diesel engine has two main aspects: variation in the speed at a certain load, and variation in the load at a certain speed of the engine. A change in the speed of the engine will also cause a change in the speed of the injection pump. Thus a higher speed will give a higher delivery of diesel fuel. A change in the load of the engine will be introduced by the position of the rack of the injection pump. Thus a higher load gives a higher delivery of diesel fuel per stroke.

In converting a diesel engine for mixed operation the diesel

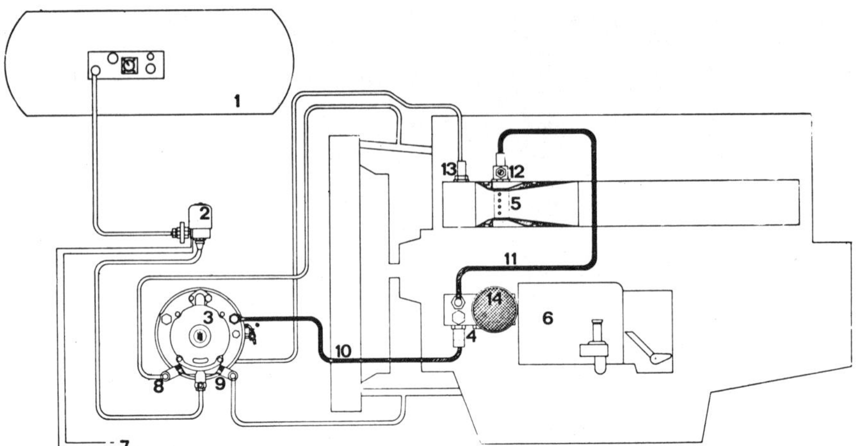

FIGURE 11. Schematic drawing of diesel-LPG or diesel-CNG mixture operation (about 35% diesel replacement, 1: fuel tank; 2: solenoid; 3: evaporator; 4: reducing valve; 5: intake venturi; 6: injection pump; 14: load corrector).

injection pump is re-adjusted to about 70% of its normal delivery. By carbureting the gaseous fuel, by means of a venturi, into the air intake manifold, the maximum power output of the engine will be restored. When the engine speed increases the amount of air passing through the venturi also increases. The increased air flow through the venturi creates a higher suction and in this way is taking a higher amount of gaseous fuel in the same time. Figure 11 shows that the gas suction line to the venturi passes through a device fitted to the injection pump. This device is called the load corrector. In the load corrector a bleed valve is mechanically connected with the rack in the injection pump.

With an increase of the load the rack comes forward and in this way also increases the flow through the bleed valve. Thus, the overall control of the amount of diesel fuel and the amount of gaseous fuel, under certain conditions, is similar.

The other hardware for the gaseous fuels - such as the storage tanks, the fuel lines, the evaporator pressure-regulator, and the solenoid valves - is exactly the same as in the conversion systems for gasoline vehicles. The air bleed system is chosen because of the low friction and because the rack of the injection pump has to travel freely to the excess fuel position for cold start.

Not only inline injection pumps are used in diesel engines, but also rotary injection pumps. The system shown before cannot be

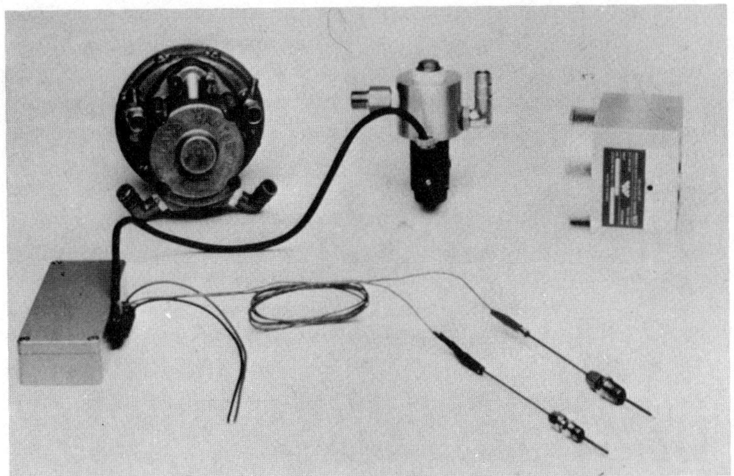

FIGURE 12. Hardware for diesel mixture operation without inline injection pumps.

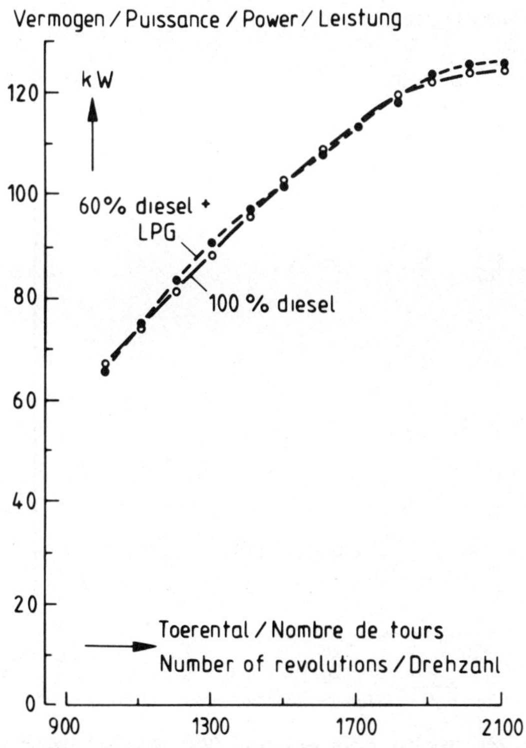

FIGURE 13. Power comparison for diesel mixture operation.

used on diesel engines with inline injection pumps. Therefore, TNO studied the possibility of designing and developing a system which can be used independently from the type of injection pump. Figure 12 shows the hardware of such a system. The evaporator pressure-regulator in the upper left corner is of a standard type. The venturi is shown in the upper right corner and in between is a proportional solenoid gas valve controlled by the electronics in the box on the left side. The exhaust temperature of the diesel engine varies with the load on the engine. By measuring the difference between air intake and exhaust gas temperature, a measure for the load on the engine is created. The temperature is measured with sensitive thermocouples and fed to the electronic control box, which can be adjusted to the engine type involved.

The maximum power output of a diesel engine operating on a mixed diesel-gas system is similar to the output of the original engine, as shown in Figure 13, although minor differences might occur.

Apart from the substitution of diesel fuel, the main advantage of the system is that the amount of black smoke is cut in half.

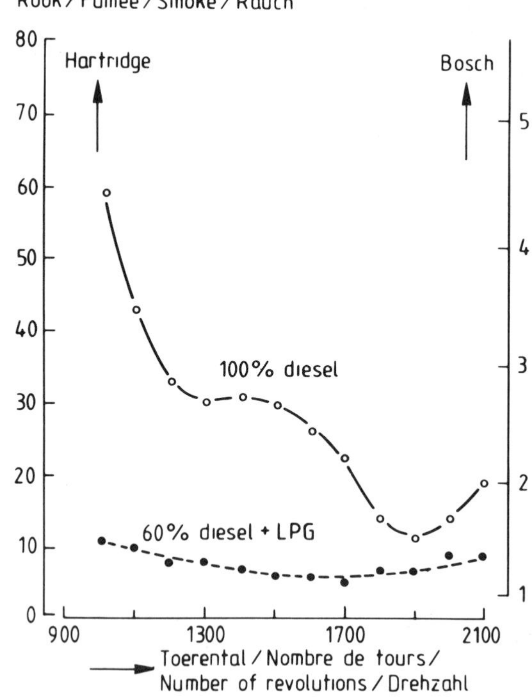

FIGURE 14. Black smoke comparison for diesel mixture operation.

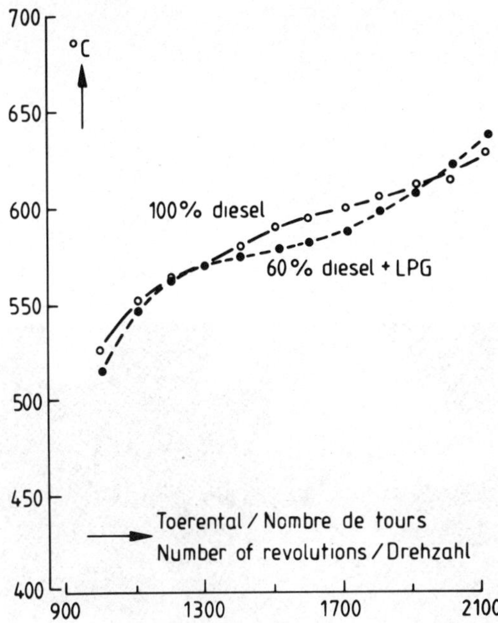

FIGURE 15. Exhaust temperature for diesel mixture operation.

Figure 14 shows the smoke curves under full load conditions of a certain engine on full diesel operation, and on mixed diesel-LPG operation. The exhaust temperature, as a measure of the engine load, is shown in Figure 15.

Safety Comments

A couple of years ago TNO conducted a large program, sponsored by Sonatrach, on crash-worthiness of LPG fueled vehicles. In Figure 16 a passenger car is crashing at a speed of 70 km/h straight into LPG tanks which are mounted, without any special protection, under the rear end of a bus. From this test program it was learned that LPG systems in collisions are as safe, or even safer, than existing gasoline and diesel oil systems.

Preliminary fire tests were also done with diesel and LPG buses, proving that, in the case of fire, LPG tanks do not necessarily explode or burst. However, a lot of research in this field has yet to be done.

FIGURE 16. Crash at 70 km/h between car and LPG fueled city bus.

THE BRAZILIAN EXPERIENCE

G. Pischinger
R. Siekmann
Volkswagen do Brasil

South America

Brazil has successfully introduced ethanol as an alternative to gasoline and the National Alcohol Program is one of the fundamentals in the energy policy of the Government. Already all the country's vehicles are operating on an up to 20 percent ethanol-gasoline blend and no problems of any significance have been experienced.

At the end of 1979, all the Brazilian passenger-car manufacturers started large-scale production of straight alcohol cars. By October 1981, nearly 7,000 alcohol filling stations were available for the 450,000 alcohol cars, which thus have nation-wide freedom of movement. The reasons for the vigorous implementation of Brazil's world-leading alcohol program, and the participation in it of Volkswagen, are dealt with in this paper.

Background of Brazil's Alcohol Program

Global Petroleum Scenario

We live in a period characterized by a very serious crisis in the supplies of liquid forms of energy; it could even lead to conflict over the vanishing stocks of petroleum. The petroleum currently produced is by no means a "conventional petroleum". Entirely new techniques for prospecting and drilling have already had to be adopted and it should be generally accepted that by the end of the century the cost of exploration and exploitation will be 10 to 20 times higher (1).

It becomes clear that the great increase in demand over the

next decades simply cannot be met by production at the current rate of increase or even by the increase obtainable by speeding up production to the highest feasible rate. There is going to be an inevitable energy shortage that will undermine the supplying of conventional fuel unless alternative energy sources are made available in the quantities required. But this "golden future" can only be reached by very intense efforts and extremely high investments. An estimate by the IIASA (International Institute of Applied Systems Analysis) in Laxenburg, near Vienna, puts the figure for the total replacement of petroleum at $40,000 billion (2).

Within this picture, road transport - as one of the main consumers of petroleum derivates - holds a key position. Thus in the larger countries such as the United States and Brazil, where people and goods are transported mainly by road vehicles, road transport accounts for more than half the country's requirements of petroleum, an energy source for which there is no real substitute in the foreseeable future. For this reason, it is imperative that we continue, as quickly as possible and with all means available, in the quest and development of fuels that are independent of petroleum. This is in the interests of the survival, not only of our current transport systems and the automotive industry as a whole, but of our very way of life today.

Brazil's Energy Situation

To accentuate better the energy scenario in Brazil, it will be contrasted with that in West Germany. Both countries put a very high priority on the development of alternative fuels. But because West Germany is a densely populated industrialized country and Brazil is still in the developing stage, with vast unpopulated areas, their approaches in reaching a larger independence of petroleum imports are quite different.

First a few facts and figures on the general situation. Next in order of size after USSR, Canada, China, and the USA, Brazil is the world's fifth country in land area. It is about 35 times bigger than West Germany and its population, at present about twice as large, is still growing rapidly, although with rising industrialization it should, around the year 2050, reach saturation at 200 million inhabitants (3). Its overall population density is only 1/17.5 that of West Germany, although in the more densely populated areas on the coast it equals West Germany. This, of course, means that the greater part of Brazil is almost uninhabited and this is the reason for the existence of many industrialization programs aimed at opening up these regions, which are rich in natural resources, and thus obtaining the two-fold advantage of reducing the rural immigration from the land and controlling the menace of the explosive growth of the large cities.

TABLE 1. COMPARATIVE DATA

1980	West Germany	Brazil	Ratio
AREA (KM^2)	249,000	8,512,000	1: 34.2
POPULATION	61.6×10^6	120×10^6	1: 1.95
PER KM^2	247.4	14.09	17.5 :1
PRIMARY ENERGY CONSUMPTION (TPE)	254.15×10^6	132.17×10^6	1.9 :1
PER CAPITA (TPE/C)	4.126	1.101	3.7 :1
GROSS NATIONAL PRODUCT (US$)	828.8×10^9	236.7×10^9	3.5 :1
PER CAPITA (US$)	13,454	1,972	6.8 :1

With regard to energy, total energy consumption in Brazil is about half that of West Germany, which puts the per-capita consumption at about a quarter. In 1980, the total energy consumed in Brazil was the equivalent of 132 million tons of petroleum, the rate of increase until then having resulted in a doubling every ten years. The West German citizen has available 4.12 tons of petroleum equivalent yearly of thermal energy, while the Brazilian has 1.10.

In 1980, the West German per capita gross national product was about seven times as much as Brazil's. Naturally the wage variations in Brazil, as in any rapidly developing country, cover a very much wider range than in a highly industrialized country like West Germany.

In absolute figures for 1980, Brazil's passenger-car fleet of 7.6 million units was still well below West Germany's 23.2 million, but the number of trucks and buses was 3.2 million in Germany against 2.4 million in Brazil (Figure 1). In population-per-car figures, still for 1980, Brazil's overall 15.8 inhabitants per car compares with West Germany's 2.7.

Of course, in the larger cities of the richer Brazilian states, the cars-per-inhabitant figure is twice the national average, while in distant regions like the State of Amazonas, the figure drops to one car for every 50 inhabitants.

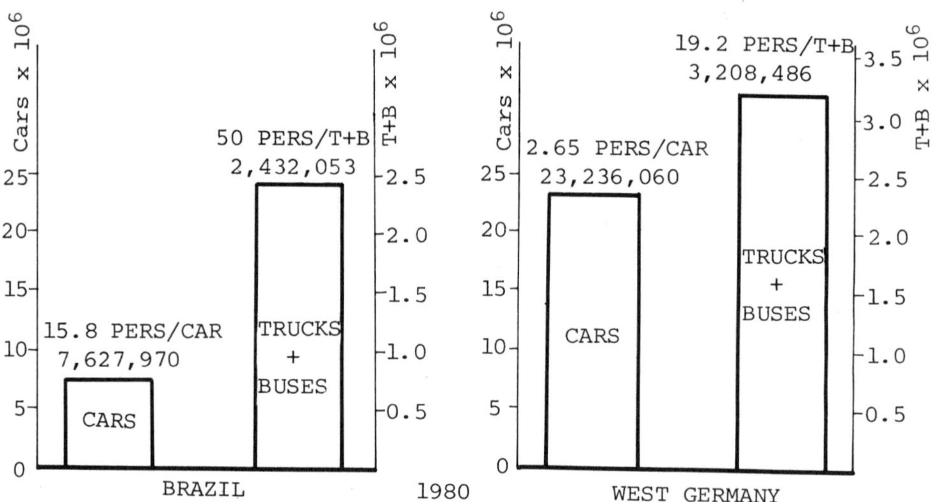

FIGURE 1. Numbers of Vehicles on the Road

Brazil's economic development has been affected by cars and roads to a much greater extent than happened in West Germany. This explains why, in Brazil today, with a situation very similar to that in the USA, more than 50% of the petroleum is absorbed by transportation, while the figure for West Germany is only about 20%.

The energy situation (Figure 2), both in Brazil and Germany, is characterized by a heavy dependence on imports. The major part of the consumption of primary energy in both countries is met by imported oil (the shaded part of the diagram indicates the proportion imported). In Brazil, petroleum's share of the energy market is 40%, of which 83% is imported, so that 33% of the country's primary energy needs depend on imported oil. Because of the explosive rise in oil prices Brazil has, in the last few years, made extreme efforts to increase its domestic production of crude, but only since 1980 have there been really positive results.

In West Germany, on the other hand, petroleum's share of the energy market is 48% of which 97% is imported.

In Brazil, water power is a key source of energy, with its substantial 26% share. Firewood, too, is significant with 20%. Coal, with 4%, is relatively unimportant; the fact that the greater part of Brazilian coal is not suitable for the production of coke means that 60% of the country's coal needs have to be imported. On the other hand, a large part of Brazil's coke requirements are met by charcoal made from wood. Sugar-cane bagasse is burnt to produce the necessary steam in all the sugar and alcohol plants.

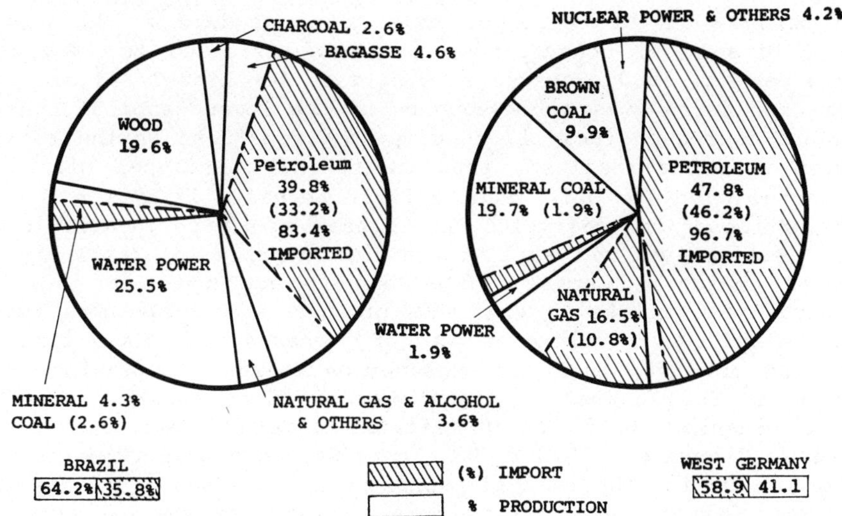

FIGURE 2. Consumption of Primary Energy (1980)

Both in Brazil and West Germany, the main petroleum derivates are gasoline, diesel oil and fuel oil (Figure 3). In Brazil in 1978, the gasoline fraction (13.8×10^9 liters/year) was lower than that of diesel oil (16.2×10^9 liters/year) while West Germany consumed about twice as much gasoline as diesel oil. In 1980, due to the energy policy of Brazil and mainly to the National Alcohol Program, the increase in gasoline consumption was halted, there

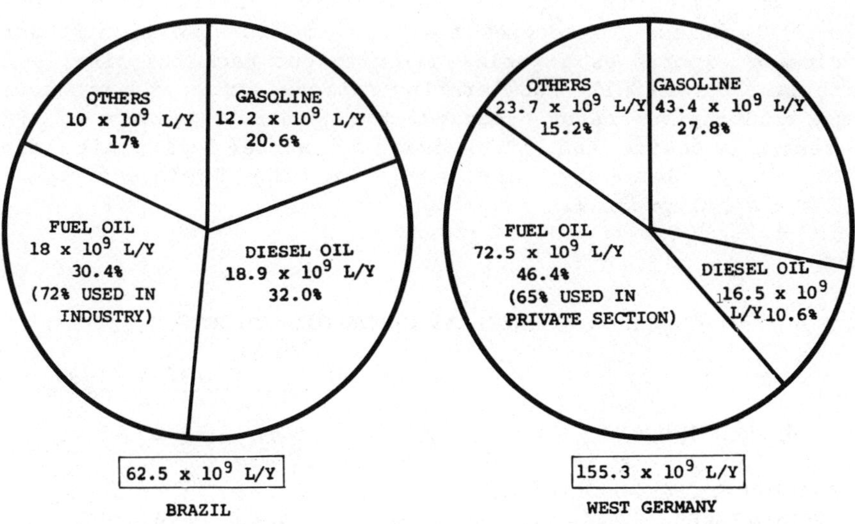

FIGURE 3. Consumption of Petroleum Derivatives (1980)

even being a slight decrease, to a total of 12.2×10^9 liters/year. Diesel oil consumption, however, increased to 18.9×10^9 liters/year in 1980, because alternatives for diesel oil are still under discussion. As a consequence of the lower gasoline consumption and increased fuel oil economy, modifications in the refinery structure can be realized, thus increasing the diesel oil output per barrel of crude oil. Fuel oil, in Germany, is by far the largest portion of the derivates and is used mainly for heating dwellings. Due to fuel oil still being low in price in Brazil, its use went on increasing until 1980, when it accounted for 30.4% of petroleum derivatives. Now, however, due to energy conservation policies and substitutions of wood and other alternatives since the beginning of 1980, fuel oil consumption shows a decreasing trend. Most of it is consumed by industries, such as cement. In 1980, Brazil consumed 62.5×10^9 liters of crude oil, while West Germany's figure of 155.3×10^9 liters was more than twice as much. In both countries the cost of petroleum imports has grown alarmingly. West Germany has to find more than double the exchange that Brazil requires for oil imports, but on the other hand, because its exports are 10 times greater, this amount corresponds to a smaller percentage in the balance of payments.

A summary of Brazilian exports and imports shows that, after the first petroleum crisis in 1973, there was a drastic increase in the share taken by petroleum in the country's total expenditure on imports. By 1978, oil imports of $4.5 billion were using up 36% of the income from all of Brazil's exports, valued at $12.6 billion.

By 1980, crude oil import expenditures already totalled about $11 billion, using up half of the country's income from exports (Table 2). Unless it becomes possible to increase significantly the value of exports as a whole and/or to cut back the rise in oil imports by substantially accelerating national crude oil production and the alternative energy programs, the country's balance of trade will remain negative and could even get worse. All this demonstrates, very obviously, the need for the implementation of Brazil's alcohol program.

TABLE 2. BRAZILIAN CRUDE OIL IMPORTS

	1978	1979	1980
CRUDE OIL IMPORTS (10^9 LITERS)	56.0	58.3	51.9
FOREIGN EXCHANGE (10^9 US$) EXPENDITURES	4.5	6.4	11
OVERALL EXPORTS (10^9 US$)	12.6	15.2	20.1

The Brazilian Pro-Alcool Alcohol Program

Brazil has a tradition in the use of alcohols as fuels for engines that goes back to the 1920's. Whenever Brazil had an over-production of sugar, this was converted into alcohol in the distilleries that usually adjoined the sugar mills. This anhydrous alcohol was then added to the gasoline in proportions that hinged on the extent of the over-production - sometimes more, sometimes less. It was only after the introduction of the National Alcohol Program in 1975, not long after the first petroleum crisis, that a national target was set for the addition of 20% anhydrous alcohol to all gasoline throughout the country. With the continuing rise in oil prices, the alcohol program was revised to allow for straight-alcohol cars and later to make a clean break between alcohol and gasoline vehicles. This has made it necessary to program considerable additional plant capacity. Generally in combined sugar mills, in addition to the production of sugar, 12 liters of alcohol are obtained by processing the molasses. By dedicating the total amount of sugar cane juice directly to alcohol production, it is possible to obtain an average of 70 liters of alcohol per ton of sugar cane (4). The Brazilian equipment industry determined that sufficient plants could be built to ensure the production, in 1985, of 10.7 billion liters a year of alcohol, of which the greater part would be hydrated alcohol with roughly 6% water content. It is considerably cheaper to produce than anhydrous alcohol but cannot be added to gasoline, as separation takes place and free water settles out. It is, on the other hand, perfectly satisfactory for its intended purpose of fueling vehicles designed to run on straight alcohol. In addition to sugar-cane as a raw material for the production of alcohol, medium and long-term proposals cover the use of manioc, sorghum and wood. While sugar-cane requires a fertile soil, both manioc and the fast-growing woods, principally eucalyptus, grow well in poor soils.

Alcohol production from sugar-cane is around 3 tons per hectare per year. No outside energy is required for the production process, not even for the distillation, because the bagasse can be used for providing process energy. These favourable properties of sugar-cane, together with distillery financing at subsidized interest rates, results in relatively low alcohol production costs of $0.32 (US) per liter anhydrous and $0.29 (US) per liter hydrated ethanol (14). In 1981, the Pro-alcool Program financed 70% of the distillery projects in the central-southern region at maximum interest rates of 55% annually and 80% of distilleries in the north-northeast at only 45% annual interest rates. This means a bargain, when compared with 95% inflation in 1981 and a projected 80% inflation rate for 1982.

The processing of manioc requires, for instance, wood as a

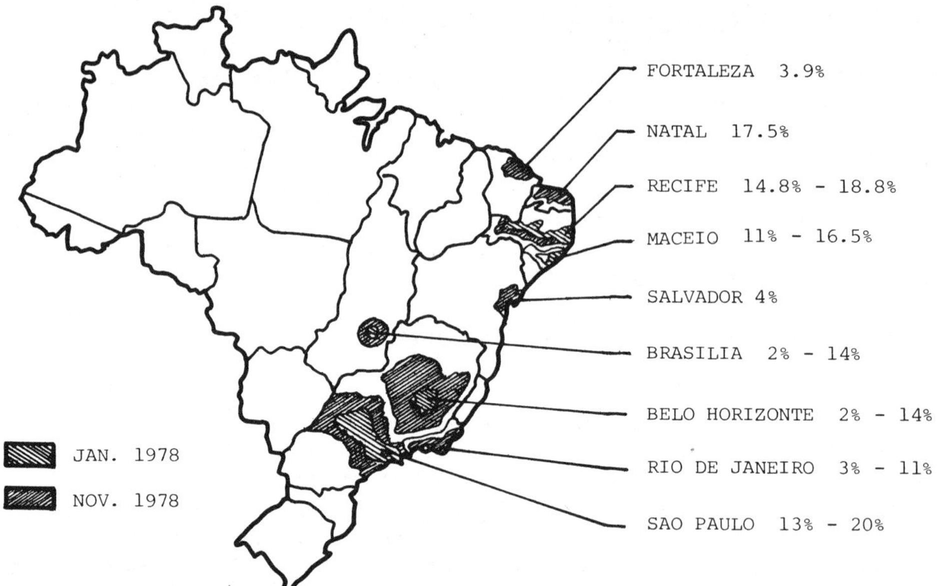

FIGURE 4. Localization and Increase of Alcohol Blends

fuel. With this proviso, alcohol production from manioc is around 3.5 tons per hectare per year. Alcohol from wood can be obtained by acid or enzymatic hydrolysis of the cellulose and hemicelluloses and may yield up to 2 tons per hectare per year.

Brazil's first manioc alcohol plants are already in production, while alcohol from wood is still in the experimental stage. Considerable improvement may be expected in all three of these alcohol-production processes - from sugar-cane, manioc, and wood - so that in the long run it may be possible to count on increases of several times the current rates.

The initial National Alcohol Program was aimed exclusively at the nation-wide addition of anhydrous ethanol to the gasoline. Figure 4 illustrates the expansion of the blends during the starting phase of the program, the blends first being used mainly in the traditional centers of ethanol production.

In January 1978, during the first stage of the National Alcohol Program, Sao Paulo State - responsible for 60% of Brazil's alcohol production - was already using a state-wide average of 13%

and, by the following November, was up to the target of 20%. Rio de Janeiro State increased from 3% to 11%, Belo Horizonte City (in Minas Gerais State) from 2% to 14%, Brasilia City also from 2% to 14%, Recife City (Pernambuco, a cane-growing state) from 15% to 19%, and Maceio City (Alagoas, also cane-growing) from 11% to 16.5%.

While the maximum of ethanol that is added to the gasoline is 20%, the actual percentage that is applied in any given locality varies with local availability. Thus there are still vast areas, mainly in the amazonian region and in the deep south of the country, where no ethanol is added due to high transportation costs. On the other hand, there are high-alcohol-production areas with relatively low local consumption where, because of distribution difficulties, part of the alcohol is used to fire boilers or is exported. The present overall national average of alcohol addition is 16%. Since the end of 1979, the Brazilian National Alcohol Program has been amplified with the going into production of straight-ethanol cars. In order to carry out this second stage, alcohol production will have to be increased, aiming at a production of 10.7 billion liters in 1985 as illustrated in Figure 5 (5).

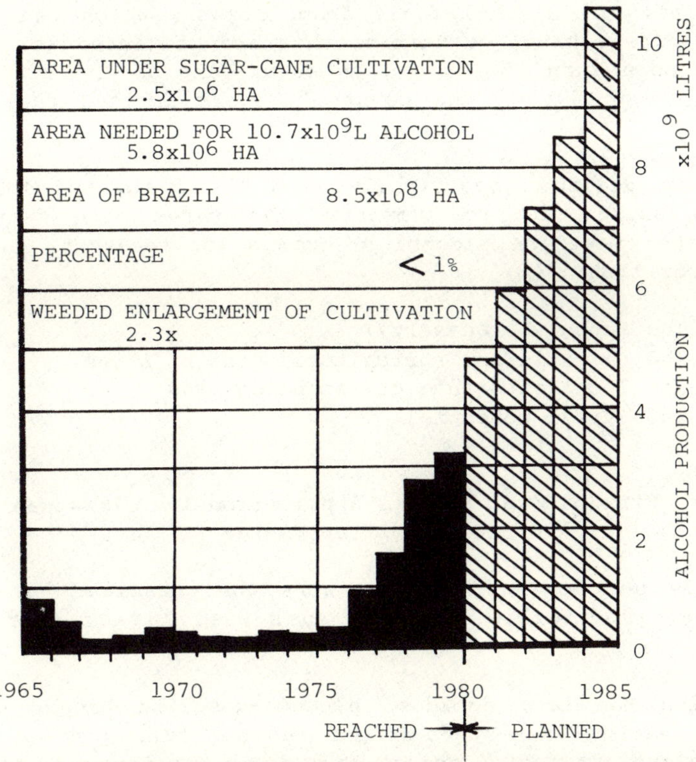

FIGURE 5. Brazil's Alcohol Program

The financing of new independent alcohol plants is in full swing. The national equipment industry is in a position to supply the installations at short notice, the time needed for installation of a new autonomous plant being between 18 and 24 months after approval of financing. What is a little more difficult to achieve is the rapid increase in the areas under cultivation, the limiting factor being the time to produce the additional cane seed required. The current area under sugar-cane cultivation in 1981 is 2.5 million hectares. Unless productivity can be improved, the area required for the production of 10.7 billion liters of alcohol and 7 million tons of sugar is 5.8 million hectares, 2.3 times the sugar-cane cultivation area of today. As a proportion of the total area currently under cultivation (40×10^6 in 1981), the total area needed for sugar-cane plantation would be only 14.5%. On the basis of Brazil's 540×10^6 ha. total area suitable for cultivation, the part to be given over to sugar-cane (5.8×10^6 ha.) would be only about 1%, so that increasing it would hardly affect the production of other crops (6).

Furthermore, the other raw materials already mentioned don't need any of the fertile land. A country like Brazil is in an almost ideal position to stage an alternative energy program based on the obtaining of alcohol fuels from biomass, which can always be renewed. The ethanol quantities that are projected for 1985 will still be more than 95%, based on sugar-cane as the raw material, and will cover 40% of the country's spark ignited fuel requirements.

Before dealing specifically with the fuels and the engines then, we would like to summarize the three main factors that between them make the alcohol program a top necessity and a feasible proposition:

1. The economic necessity.
2. The climatic and agricultural possibilities.
3. The local existence of technology and industrial potential.

Ethanol Fuel and its Application in Volkswagen 1.5 Liter Engine

It is not for technical reasons that gasoline has been, up until now, practically the only fuel used in our spark ignited engines.

During the last decades, alcohol-gasoline blends have been used for various moment-to-moment reasons but, just as they appeared on the automotive scene, they again disappeared, as the particular reason for their use had ceased to be valid. As shown before in this paper, we once more have the motivation for considering alternatives to crude oil derivates, but this time we believe that it has a more dramatic background: we can foresee the end of the supply of crude oil at least in terms of the increasing volumes required at a reasonable price. So, once again, alcohols are being discussed intensely as an alternative to gasoline derived from crude oil. The specific properties of ethanol and methanol make them potentially excellent spark ignited engine fuels. In a study by the Union Oil Company of California (7), the preference between these two is very clearly for ethanol, both for blending and as a straight fuel. Nevertheless, the use of ethanol in Brazil within the National Alcohol Program was not determined by studies but by the already mentioned facts:

- previous experience of its application to vehicles (since 1923);
- ethanol production facilities already existing; and
- favorable climate and terrain for growing the raw materials (biomass).

Considerations on Ethanol/Gasoline Blends and Straight Ethanol

As mentioned before, a 20% ethanol blend of gasoline has already become a familiar fuel to millions of Brazilians in everyday life without any significant problems. There are no indications that the lifetime of engines originally designed for gasoline is affected negatively although, in the fuel system of some models, some plastic and/or rubber materials had to be changed. Emissions from non-emission controlled vehicles using ethanol-gasoline blends are in general better (8).

The total amount of gasoline the country can save by blending ethanol into gasoline for use in straight-gasoline cars is limited by the proportion above which drivability is adversely affected. For the Brazilian car population, this is 20%. Use of a higher-ethanol blend would not make sense as it would entail engine modifications similar in kind and cost to those for running on straight ethanol. Therefore, the only logical long-run step to further substitution of gasoline is the introduction of straight-ethanol technology, as was done in the second stage in the National Alcohol Program in Brazil.

One has to be well aware that by this the fuel's compatibility with today's gasoline cars, as still exists in the case of the ethanol-gasoline blend, was dropped. A new generation of straight-ethanol cars was required as well as additional pumps and tanks at the filling stations. This means that, especially during the implantation of the straight-ethanol technology, tough and careful

planning and control is required to have an adequate number of ethanol cars, an adequate and well distributed filling station network and an adequate production of ethanol. A very sensible and critical part within this system is the purchaser of the ethanol cars. If he fails to buy, even the best program is bound to fail.

Therefore, it is indispensable to generate confidence and to promote the new technology by incentives. In Brazil the incentives granted by government are as follows:

- ethanol's price at the pump is not more than 65% of the gasoline price. (Although this reflects exactly the amounts of energy per liter, it is still an incentive when considering the roughly 20% better efficiency of the ethanol engines.)
- road tax for ethanol cars is half of that for gasoline cars.
- gasoline filling facilities are closed from Friday evening through Monday morning, whereas ethanol is available also on Saturdays.

The difficulty in controlling such a substitution program when the alternative fuel is not compatible with today's gasoline car-population may be shown by the set back in the beginning of 1981. A few notices in the press appeared, announcing that there might be a temporary shortage of ethanol, which, in fact, never occurred. As a result, purchases of ethanol cars declined sharply and recovery was slow.

Automotive Industries as Partners in the National Alcohol Program

Having produced 923,627 cars in 1978, Brazil ranked tenth amongst the 28 car-producing nations (9). Brazilian cars are assembled almost exclusively with nationally produced components. Volkswagen do Brazil's cars, for instance, are 99% national by weight. Therefore Brazil fulfills one of the most essential requirements for an alternative fuel program: a national automotive industry capable of producing cars to national requirements and having a vital interest in doing so. On September 19, 1979, a contract was signed between the automotive manufacturers' association (ANFAVEA) and the Brazilian Government (10), for the production of the following numbers of alcohol cars:

1980	250,000
1981	300,000
1982	350,000

At the beginning of 1980, all of Brazil's car manufacturers had ethanol versions of some of their models in regular production and, since March of 1980, they have been available to the public and not only, as before, to governmental and semi-governmental entities. By October 1981, there were about 450,000 cars operating on straight ethanol, which is available at nearly 7,000 filling stations all over the country, giving nation-wide freedom of movement to these ethanol cars.

Brazilian Straight-Ethanol as a Fuel

For investment and production-cost reasons, it has been determined that hydrated ethanol will be used to meet:

- Low-cost production, and
- Lower investments for new distilleries.

With the average technological development of Brazil's existing distilleries in mind, ethanol quality was set by the specifications in Table 3.

TABLE 3. SPECIFICATIONS FOR ETHANOL

CHARACTERISTICS	HYDRATED ETHANOL AS AUTOMOTIVE FUEL
SPECIFIC GRAVITY AT 20°C	0.8073 - 0.8150
ALCOHOL CONTENT WT %	91.1 - 93.9
RESIDUES, MG/100 ML. MAX.	5.0
TOTAL ACIDITY, MG/100 ML. MAX.	3.0
ALDEHYDES, MG/100 ML. MAX.	6.0
ESTERS, MG/100 ML. MAX.	8.0
SUPER ALCOHOLS, MG/100 ML. MAX.	6.0
COPPER, PPM. MAX.	NOT SPEC.
ALKALINITY	NEGATIVE
APPEARANCE	CLEAR AND WITHOUT MATERIAL IN SUSPENSION

Unfortunately, the temperature control systems for the batch fermentation process are, in many existing plants, insufficient. The resulting higher aldehyde contents become acids by oxidation, so the specification had to take this into account and allow a relatively high acid level. However, many new plants are being set up and in these there is a marked tendency towards product improvement.

Properties of Ethanol Compared With Gasoline

The properties are shown in Table 4. The differences in boiling temperatures, heat value, heat of vaporization and stoichiometric air/fuel ratio for combustion are responsible for most of the problems encountered when using straight ethanol as spark ignited engine fuel. But advantages, such as higher octane numbers and a better ignitability of lean mixtures, have also to be discussed for the better understanding of engine modifications and results.

TABLE 4. PROPERTIES OF HYDRATED ETHANOL AND REGULAR GASOLINE IN BRAZIL

		ETHANOL	REGULAR GASOLINE
WATER CONTENT	WT %	~ 6	-
DENSITY AT 15°C	G/CM^3	0.8	0.735
BOILING TEMPERATURE	°C	78	32 - 185
HEATING VALUE	LOWER KJ/KG	~ 25140	~ 42430
LATENT HEAT OF VAPORIZATION	KJ/KG	903	376 - 502
STOICHIOMETRIC A/F	KG AIR / KG FUEL	8.45	14.8
MOTOR OCTANE NUMBER	MON	94	73
RESEARCH OCTANE NUMBER	RON	111	85

In theory, the heat value of hydrated ethanol causes a roughly 54% higher volumetric consumption but, due to the stoichiometric air/ethanol ratio, the heat value of the mixture remains comparable with the stoichiometric air-gasoline mixture. The latter fact guarantees at least a comparable power to displacement ratio which means, in other words, that one can have the same engine size to move the same vehicle.

The boiling temperature and the heat of vaporization of the alcohols hamper the mixture formation, particularly when carburetors are used. In that case, to obtain good drivability and good quick throttle response along with lean mixtures for low consumption and emissions, the mixture has to be heated substantially before entering the cylinders. Also, at below + 15°C cold starting of the engine is not possible without auxiliary systems. The high octane numbers allow a significant improvement in the thermal efficiency by permitting increased compression ratios. Fuel-efficient lean carburetor settings were obtained without any problems in drivability, so it may be concluded that a well-prepared ethanol-air mixture offers better ignitability than gasoline-air mixtures in the lean range.

Volkswagen do Brazil's 1.5 Liter Water-Cooled Ethanol Engine

For the engine lay-out, we take the example of this engine because it belongs to the same family that powers the Rabbit, Dasher and Audi 4000.

Technical Data (gasoline engine figures in brackets)

Cylinders	4 in line
Bore	76.5 mm
Stroke	80.0 mm
Displacement	1471 cm^3
Compression ratio	10.5:1 (7.5:1)
Valve gear	OHC
Carburetor	one-barrel, downdraft, manual choke
Cooling system	water-cooled, two circuit with thermostatic control, water pump, radiator with electric fan
Ignition system	Coil ignition with high energy output
Spark-plug gap	0.7 mm

Objectives

- Minimized consumption
- Performance comparable with the gasoline version
- Comparable drivability
- Comparable durability
- Same intervals of maintenance
- Same serviceability
- Low costs.

In order to obtain the same serviceability, it was decided to use only the technology familiar to the ordinary, VW service personnel in Brazil. We therefore based our work on the single-stage carburetor and the conventional ignition system of the then-marketed gasoline version. In our 1982 models we introduced electronic ignition in both the ethanol and gasoline versions.

Engine Lay-out

For our ethanol engine, we adopted a compression ratio of 10.5:1. Theoretically it could have been higher, but we wanted to avoid risks to the durability and we also had to consider knocking at full load due to pre-heating of the mixture. Furthermore, increasing the compression ratio above 10.5:1 would not have produced any considerable increase in thermal efficiency.

Figure 6 shows that, with a compression ratio increase from 10.5 to 12, the improvement in consumption would be around 2% at 60 km/h, and around 3% at 100 km/h.

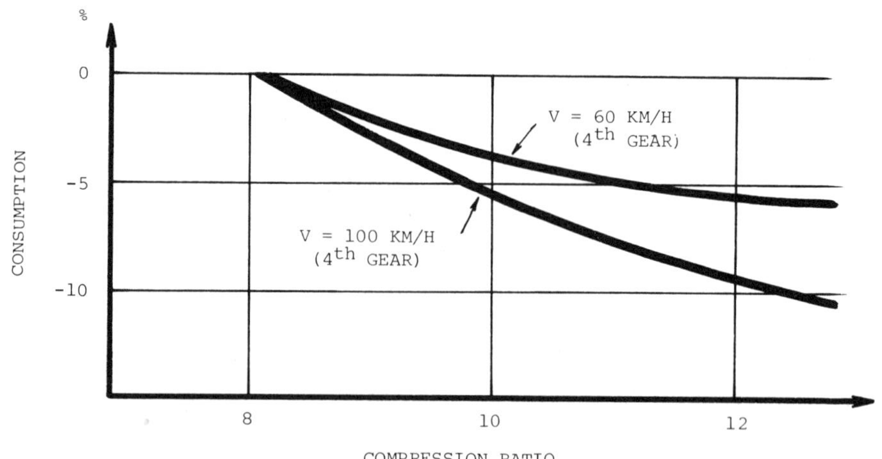

FIGURE 6. Gain in Part-load Fuel Consumption vs Ratio in Ethanol Engines (11)

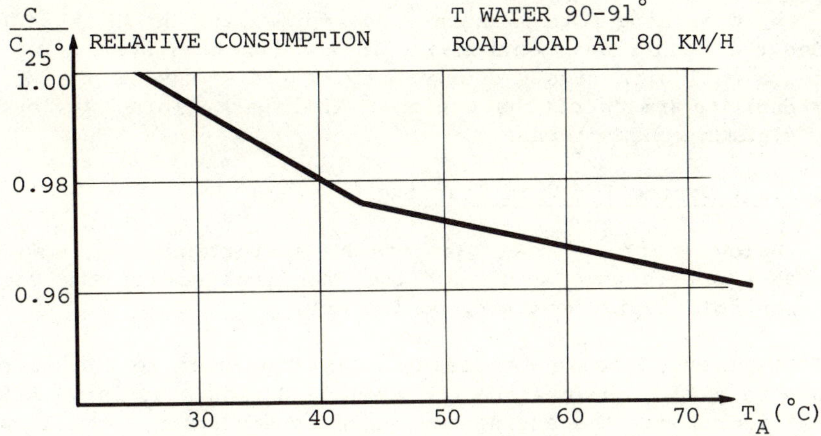

FIGURE 7. Influence of Intake-Air Temperature on Consumption

To preheat the mixture before it enters the combustion chamber, we apply hot intake air (up to 90°C on entering the carburetor) and heat the walls of the intake manifold intensely. The intake air is forced to pass over the exhaust manifold before entering the air filter. The intake manifold gets heated by a water jacket covering as much area as possible. The jacket is connected to the cooling system, which thus feeds it with the hot water coming directly from the cylinder head. Figure 7 shows the effect of preheated intake air on the minimum-obtainable specific fuel consumption at a constant speed of 80 km/h (measured on the test bench). A gain in economy of nearly 4% was obtained.

Figure 8 shows the influence of heating the walls of the manifold under the same conditions, the temperature of the intake air being 44°C.

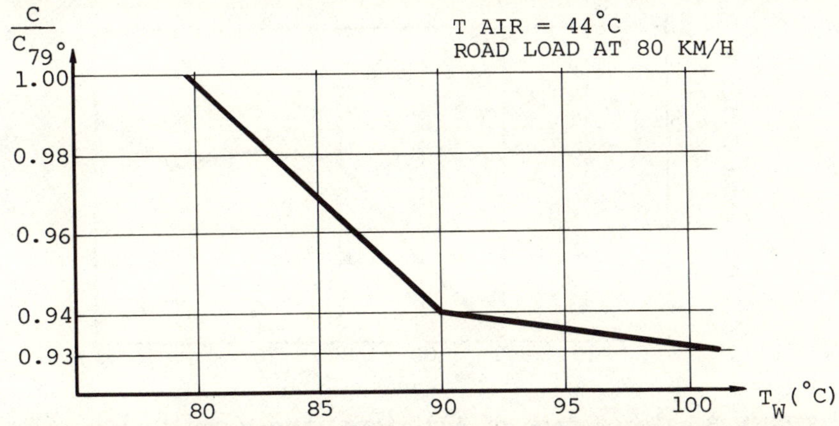

FIGURE 8. Influence on Consumption of Water Temperature in Intake Manifold Heating Jacket

By combining both measures, an improvement of 9% is achievable under the conditions mentioned. In the ignition system, we introduced a coil with higher energy output. This gave better drivability during warm-up of the engine. The spark timing was optimized for minimum consumption.

Auxiliary System for Cold Starting

Below + 15°C we use gasoline as a starting fuel. We do not use an electric pump to inject the gasoline; we let the vacuum in the manifold during starting do the job.

A solenoid valve is electrically connected to the starter in order to permit the passage of gasoline during starting. After the first ignitions, the engine continues on ethanol. A small gasoline tank (about 2 liters) is installed in a place which was found to be safe in crash tests.

Power and Torque

The development was focussed on minimum-obtainable fuel consumption rather than on high power and torque outputs.

Net Power:	47.8 kW/5600 rpm (DIN)
	(47.8 kW/5600 rpm (DIN)
Net Torque:	107.9 Nm/3000 rpm (DIN)
	(101 Nm/3000 rpm (DIN)

(values in parenthesis are for the Brazilian gasoline version).

Consumption and Emissions

The consumption under road load conditions (4th gear) is shown in Figure 9.

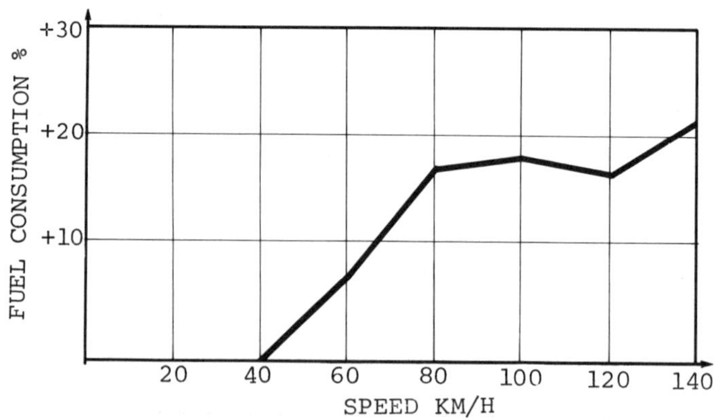

FIGURE 9. Percentage of Additional Fuel Consumption in Relation to a Similar Ethanol-Gasoline Vehicle (4th Gear, Road Load)

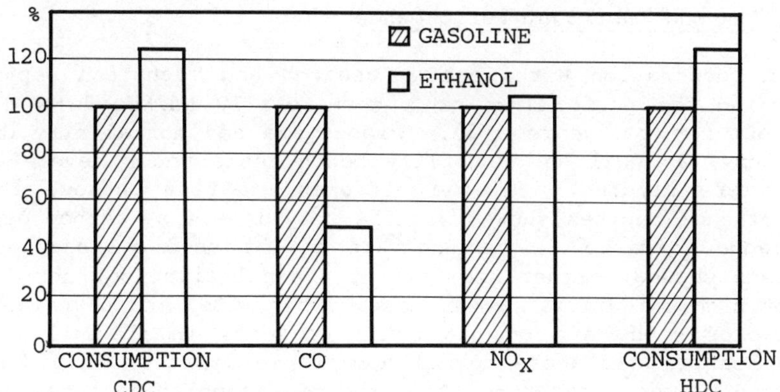

FIGURE 10. Comparison of Ethanol and Gasoline Engines. (Consumption in CVS City Driving (CDC) and Highway Driving Cycle (HDC) -- Exhaust Emissions in CDC)

The increase in volumetric consumption over the gasoline version hardly exceeds 20%. This result has been confirmed in the CVS test, where the difference was found to be roughly 22%. The results are shown in Figure 10.

Both are non-emission controlled Brazilian engines. The theoretically higher consumption of roughly 54%, stemming from the properties of ethanol, is substantially diminished.

CO being 45% lower with the ethanol version indicates comparatively lean operation. NO_x being slightly higher reflects both the higher compression ratio and the leaner operation as well as the permanent preheating of the mixture.

Data on HC is not reliable when measured normally with CVS equipment (12, p. 172).

The total aldehyde emissions are six times higher than those of the gasoline car. Measurements with gas-chromatography showed a ratio of 3:1 between acetaldehyde and formaldehyde emissions in the CVS-City driving cycle. Future research work will have to show whether a reduction of these emissions is necessary and how this can be obtained.

Drivability

Due to heating of the mixture, drivability is totally normal with the engine warmed up. After a Brazilian winter cold start at + 10°C, it takes about two to three minutes to reach good drivability on a still slightly rich mixture.

Durability and Lubricant Performance

In cooperation with Texaco Research and Technical Department we studied the performance of a commercially marketed SE quality motor oil for two years (13). High-speed and normal city driving conditions, as well as durability bench tests under severe conditions, were studied. In the different testing phases, the 1.5 liter ethanol engines showed satisfactory levels of carbon deposits and sludge formation. The analysis of the used oil samples from the tests showed neither significant ethanol dilution, nor indication of high oxidation. Total base number (TBN) and pH values were satisfactory and data on the metal contents in the oil gave no indication of wear above normal acceptable limits. This last was confirmed by wear measurements of the disassembled engines, showing that wear of the ethanol-fueled engines tested remained within the VW specifications fixed for gasoline engines under similar conditions.

With respect to upper-engine lubrication, ethanol behaves like lead-free gasoline, so it was necessary to apply the same modifications to the engine as are required for lead-free gasoline in the USA. This involves mainly valve stems (chromium-plated valve stems with induction tempered ends) and the seating surfaces of the exhaust valves.

Engine tests by the usual VW procedures before going into production showed durability entirely comparable with that of the gasoline version.

Compatibility of Materials

Considerable problems can be encountered with some materials when they come into contact with ethanol. Results of a large study are given in Reference 7.

Brazilian ethanol offers somewhat more severe conditions. The high acid content and the presence of water give rise to additional problems. Table 5 shows the weight loss of Zamac plates in a seven-day corrosion test in ethanol at 50°C. The framed lines give the results of the effect of increased acid levels at a constant water level. Increasing the water level along with the higher acid content has a tremendously strong influence. It may therefore be concluded that water and acid content, or both, must be kept as low as possible.

Values in brackets are the weight loss when using 0.25% of a commercial corrosion inhibitor.

TABLE 5. INFLUENCE OF ACID AND WATER, PRESENT IN ETHANOL ON THE CORROSION OF ZAMAC

ACETIC ACID (MG/100ML)	WATER CONTENT (WT %)	WEIGHT LOSS (G/M)
1.5	6.2	1.1
1.5	9.0	1.1
1.5	11.7	3.6
3.0	6.2	2.0
3.0	8.9	5.0 (0.1)
3.0	12.0	28.0 (4.0)
3.0	15.0	42.5 (21.0)
5.0	6.2	8.3

We give a few examples of difficulties encountered: zamac (zinc/aluminum alloy) suffers heavy corrosion, with the formation of oxides and hydroxides of zinc. We apply a two-stage bath to coat carburetors and fuel pumps made of zamac. The first stage is a treatment to provide better adhesion of the second impervious layer (dichromatization).

Brass and steel in the vicinity of zamac are subject to electrolytic pitting supported by the presence of water in the ethanol, so they are either replaced by plastic or undergo the same treatment as zamac.

The sheet-steel fuel tank is coated, on the inside, with plastic. If, however, there should be any slight porosity in the protective coatings of the fuel tank or of the carburetor, the occurrence, under extreme conditions, of some corrosion cannot be entirely excluded, so we recommend adding corrosion inhibitors to the ethanol when filling the tank. Such inhibitors have been approved in tests and are commercially marketed in Brazil.

Future Ethanol Fuels and Engines

For the time being, as we have said, our fuel is simple hydrated ethanol, but in the future we should have a more complete spark ignited engine fuel based on ethanol.

Corrosion Inhibitors

It is certainly cheaper to protect the existing chain of

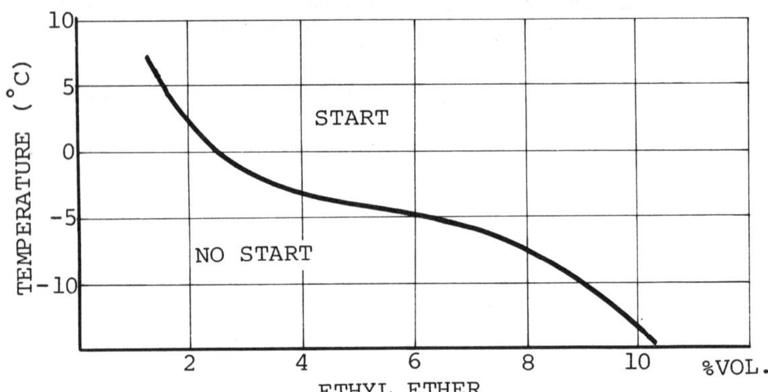

FIGURE 11. Influence of the Addition of Low-boiling Ethyl Ether on Start Temperature Limits of Ethanol Engines

supply against corrosion by adding corrosion inhibitors to the fuel than to rebuild it with other materials.

Several additives are being studied, some for retail use being already on the market, but up to now there has been no general use of anti-corrosion additives.

ANFAVEA, the Brazilian association of automotive manufacturers, has a group of experts working in this field with a view to coming up, as soon as possible, with a fuel specification that includes inhibitors.

Vaporization Characteristics Improvement

The auxiliary cold-start system can be eliminated by adding a low-boiling point additive. Several suggestions are given in Reference 11.

For the Brazilian climate, with temperatures never falling below - 10°C, we suggest an addition of ethyl ether, which has a boiling point of 34.5°C and is a derivate of ethanol.

Figure 11 shows the start temperature limits versus the addition of ethyl ether. A 10 % addition would provide a sure cold-start at -10°C.

Also to be considered is the supplying of some regions, during the winter season, with a winter-type fuel, as is done with diesel fuel. The addition would, at the same time, substantially improve drivability and consumption during warm-up period. When considering the ethanol technology for countries with cold climates, such additives should certainly be in the fuel.

Considerations on Future Ethanol Engines

In order to improve fuel consumption and emissions, fuel injection and electronic control systems will be applied as well as all the other measures now being considered with a view to improving the economy of today's gasoline vehicles.

Beyond this, however, attention must be given to ethanol's very particular properties. The latent heat of vaporization of the stoichiometric air/ethanol mixture is four to five times higher than of the stoichiometric air/gasoline mixture.

When using fuel injection (with lower requirements in preheating of the mixture), the volumetric efficiency and thus power and torque at WOT can be increased. Either smaller engines or gear ratios with overdrive characteristics can be applied for optimization of the engine-vehicle system for minimum consumption.

Even more advantages will be encountered when supercharging the engine. An adiabatic vaporization of all ethanol in a stoichiometric mixture gives a temperature decrease of about 125°C. Therefore compression ratios of turbocharged engines can be higher, thus improving consumption in city driving and providing a higher power output. Again, smaller engines can be applied to increase economy. Last, but not least, dissociation of ethanol before it enters the combustion chamber may be considered as a further potential for improving ethanol-engine technology.

Brazil has taken a great forward step in the use of straight ethanol as a spark ignited engine fuel, with economic necessity and the country's resources as the stimuli.

In the worldwide search for alternatives to gasoline, Brazil's experience is already being sought by other countries interested in introducing similar programs. It may be that, for some countries, fuel other than ethanol might appear more feasible, but in the interest of maintaining the international exchange of cars and of facilitating border-crossing traffic, every attempt should be made to have similar alternative technologies around the world with a high degree of compatibility between them.

REFERENCES

1. Statement during opening session by Shell, Proceedings of the X World Petroleum Congress, October 1979, Bucharest, Romania.

2. Volkmann, D.J., "Der Kapitaleinsatz beim Uebergang zu neunen Energieversorgungssystemen", Energiewirtschaftliche Tagesfragen, May/June 1979; Bonn, West Germany.

3. Fucks, W., Maechte von Morgen, Deutsche Verlagsanstalt, Stuttgart, 1978, West Germany, chapter Kraftfelder, Tendenzen, Konsequenzen.

4. Heitland, H.H., et al, "Use of Ethanol from Biomass as an Alternative Fuel in Brazil", Proceedings of the International Symposium on Alcohol Fuel Technology, Wolfsburg, West Germany, Paper 1-3, 1977.

5. Modelo Energetico Brasileiro, Ministerio das Minas e Energia, Republica Federativa do Brasil, November 1979, Brasilia, Brazil.

6. Communication EMBRAPA, July 9, 1981.

7. Nagaguchi, G.M., Keller, J.L., "Ethanol Fuel Modification for Highway Vehicle Use", Final Report, June 1979, Union Oil Company of California.

8. Pischinger, G.H. and Pinto, N.L.M., "Experiences with the Utilization of Ethanol/Gasoline Blends and Pure Ethanol in Brazilian Passenger Cars", Proceedings of the Third International Symposium on Alcohol Fuel Technology, Asilomar, Cal., Vol. 1, 1979.

9. World Automotive Market, Johnston International, 49th Edition, 1979.

10. "Protocolo do Alcool", Associacao Nacional dos Fabricantes de Veiculos Automotores "ANFAVEA", Circular no 031/79, September 21, 1979.

11. Menrad, H., "Aethanol als Kraftstoff der Ottomotoren", ATZ 81 (1979) 6, Franck'sche Verlagshandlung, Stuttgart, pp. 279.

12. Heitland, H.H., Bernhardt, W., Lee, W., chapters 2.1/2.3 in Neuen Kraftstoffen auf der Spur, Bundesministerium fuer Forschung und Technologie, Bonn 1974, West Germany.

13. Pischinger, G., Kilby, D., et al, "Lubrication of SI Ethanol Engines", Proceedings of the Fourth International Symposium on Alcohol Fuels Technology, Guaruja' - SP - Brazil, 1980.

14. Proceedings of the First Congresso de Alcoolquimica, Sao Paulo, June 23-26, 1981.

ASPECTS OF THE ALTERNATIVE FUEL PROGRAM IN THE UNITED STATES

Dan Glickman

Congressman
United States House of Representatives
Washington, D.C.

Canada has shown great leadership in a variety of energy alternatives, as have many countries of the world, including my own country. I would like to give a little bit of background about where the United States is in terms of the future of alternative energy sources, particularly methane.

The United States is the largest user of energy in the world and in terms of our productive effort, although certainly we no longer lead the world in direct oil production, we do lead indirectly because it is through all of our many companies, located abroad and headquartered in the United States, that most of the world's oil production is actually made. I am amused that there are three or four politicians leading off this conference when in fact so many of our problems may be due to our activities. Are we in government going to be willing to try to articulate better national energy policies so that the public does not believe that all we are spreading is confusion in the world? That problem is of immediate concern.

An important foreign policy issue facing the United States today is whether the U.S. should sell AWACS aircraft to Saudi Arabia. AWACS, as you probably know, is basically a 707 that has a round saucer on the top which rotates. This round saucer is capable of enormous, complex, and sophisticated aerial intelligence capabilities and the Israeli Government naturally isn't interested in having it sold to Saudi Arabia. The question is, should we or should we not sell AWACS to Saudi Arabia?

The underlying issue is that as the U.S. still imports roughly 35 to 37% of its oil from Saudi Arabia and related countries, we may run the risk of being cut off if we don't sell them the AWACS. Maybe not tomorrow, but perhaps next year, and obviously that key in foreign policy is critical in almost every decision that the U.S. makes. That is why the whole issue of finding an alternative to OPEC is so critical.

My own interest in methane actually got started in a couple of areas. I represent Wichita, Kansas, and that general area, in the U.S. House of Representatives. We produce about 60% of all airplanes in the world in my home town. There are small aircraft companies and one of them, Beech Aircraft, some time ago became interested in liquified natural gas as a motor fuel and brought it to my attention. Another company located in Wichita, Advanced Fuel Systems, raised the same subject with me and the American Gas Association. My interest in Kansas also stems from the fact that in the Mid-West we are large producers of natural gas. With all that in mind, in the last Congress I introduced the Methane Transportation Research Development and Demonstration Act. This legislation will do roughly what is being done in Canada. We would provide $13 million over a five year period to do some basic research but the key part of the program was to do demonstration programs primarily in fleet use for compressed and liquid natural gas.

When I introduced the bill, there was a lot of hositility to it but even more questions about it. It took some pulling of teeth but we were able to pass that legislation in December of 1980. It was one of the last bills that President Carter signed and it went into effect on October 1.

Unfortunately, however, before the program even got moving there was talk from the new administration of not funding it and, in fact, repealing almost every piece of alternative energy legislation that has been passed in the U.S. in the past three or four years. Fortunately it has not been repealed and there will be alternative transportation fuels funding that our Energy Department can put into methane research.

Most of the monies now are going into basic R & D related activities and I don't think the demonstration program is going to be funded. In fact, President Reagan has indicated his interest in abolishing the entire Department of Energy! The President and I are not of the same political party and we disagree on most things. This is another issue we happen not to agree entirely on. I think it is a foolish decision. I recognize some of the regulatory problems this Department has had but I think the signal to the world of abolishing it is the wrong signal to give the world when in fact energy is probably the main foreign policy and economic problem facing the U.S. It is not a terribly productive move in terms of our own energy self-sufficiency and it certainly is not productive for the energy self-sufficiency of North America.

I recognize that the private sector in the U.S., as well as in most countries, will take the lead in developing technology but I happen to believe that abolishing the Energy Department could diminish that initiative.

I know that with respect to the program in the U.S. the Energy Department has commissioned an assessment of gaseous fuel in automobiles. The first "official" draft report came out in September and we are lucky that it looks nothing like an earlier "unofficial" draft. The earlier version went out of its way to dream up hypothetical problems that on-the-road experience elsewhere showed to be ludicrous. However, this new report outlines issues that need research attention: fuel economy, lighter-weight tanks, potential for turbocharging, and use in diesel engines to name a few. Given the negative bias in that earlier draft, I will be giving this "first draft" careful scrutiny: the next major goal is to get the wheels moving so those research needs will be met. If there is follow through on this some meaningful advances could be made, primarily in research.

With the on-the-road experience, particularly in Italy and to a limited degree even in the United States, all of this might sound like a waste of effort. But before we're going to see investors move full speed ahead into methane-fueled vehicles, they will need to know that government safety, environmental and fuel-economy standards won't create problems down the road. Front-end research and testing can clear the way at the same time it enhances this technology, and I thought I would mention a few other things that are happening in the U.S. that may affect methane fuel.

First off, there is the price of natural gas. In 1978 the U.S. Government passed legislation that would partially deregulate the price of natural gas which, as many of you know, had been controlled artificially since the 1950's far below the Btu equivalent of other energy - and particularly petroleum. There is, as a result of the fact, some natural gas totally decontrolled and new natural gas, in some cases in very deep natural wells. Much, though, is only now being gradually decontrolled and there is the potential for an enormous price increase depending on what the world market situation is in 1985 when decontrol takes effect.

Some in the administration and some in Congress would like to see natural gas immediately decontrolled. There are forces, including a lot of people in the utility industry, who believe that might be a mistake for a couple of reasons: firstly, because of problems in demand and, secondly, because of the inflation problems in the U.S. My prediction is that we will not immediately decontrol natural gas but that it will continue in its current, what I call "phase", deregulatory stage. While that may have some economic problems to it, it is simply a fact of life and, on the practical side, the increasing prices for both crude oil and natural gas in the U.S. have led to an unbelievable increase in drilling activity. This can be combined with the fact that the windfall profits tax has been repealed for all independent producers in the U.S. With some exceptions, therefore, I think there are enormous incentives for the private sector to develop natural gas in the future.

The second point I would like to mention is the issue of highway taxes with reduced consumption of petroleum in the United States. I am sure this is true in Canada and the rest of the world. We have a horrendous problem with maintenance of highways and yet our federal highway tax in the U.S. which is now four cents a gallon, has been roughly the same for 20 years. President Carter tried to raise it and it was voted out then by a ten to one margin in the Congress. With an increased use of alternative energy sources, that situation clearly is going to have to be dealt with by policy makers. It's an important issue that just can't be forgotten when we move away from petroleum.

The third thing I would like to address is the automobile companies. Dr. McGeer mentioned that, with some exceptions, they are not represented at this conference. I believe that I saw a Ford Motor representative here and perhaps one from Chrysler. But General Motors is not here. I visited the General Motors Research Center in Detroit. As I understand, it is the largest private research entity in the world - in terms of the number of folks working there - that is not strictly set up as a private sector research think tank. Close to 10,000 people work there. When I went to visit them in Detroit, Michigan I asked about their work on methane because obviously General Motors is a leading producer of cars in North America. I got the impression of a couple of things: one is that they are not doing tremendous work in methane right now; they're doing a lot of work in pulverized coal, but their alternative energy work is not very significant compared to the work that they are doing on catalytic converters, exhaust systems, improving and fine-tuning the internal combustion engine, and in diesel work. I asked why, and the general response was that while there is additional work to be done on methane, they do the research problems in available fuel, transmission systems, transportation and things not related to the technical aspects of building a methane car. And I suppose that our automobile companies have certainly fought the development of fuel-efficient vehicles every step of the way and have brought a lot of their own troubles with them.

I do not believe that they will be the ones that will develop methane-powered cars on their own. I think it is going to be either smaller companies or it is going to be on the demand side. It may be utility companies or it may be fleet operators who begin to demand the kind of fuels that will be less expensive, safer, higher in octane, and with less environmental problems that will push the auto companies to begin developing theirs.

What I am saying is they are not going to do it out of the goodness of their hearts and I doubt that the Government of the United States is going to force them to do it. Therefore, the market will have to push them along. I'm hopeful about that, but the thing that worries me is friends I have in the government who say the market needs to push them along. If the government loses sight of the key role in policy and in symbolism, then General Motors, Ford, Chrysler and perhaps some of the German and Japanese auto companies, will lag far behind. I think the biggest challenge we face in the U.S. is the government that abdicates any responsibility whatsoever for energy involvement and lets it all go to the private sector.

I don't think that we have to reinvent the wheel, or should I say rediscover methane potential, in country after country. I hate to see things like the grant to the B.C. Research Council happen everywhere in the world so that the technology work is being duplicated in 57 different countries. I hope that we can move ahead right away and share a lot of the reseach of the potential that is available.

There is no question at this stage that the internal combustion engine is basic to all our transportation systems. We all have domestic access to methane from coal, natural gas, sewage and other wastes. At least one scientist, Dr. Thomas Gold of Cornell University, is convinced that methane is a major component of the earth's core. What more international an energy resource could we find to learn how to develop together?

Unfortunately, energy has tended to be a divisive, rather than a unifying force between nations. Even traditional, long-standing friendships have been strained by disagreements over fuel. The current jousting between the Canadian Government and my own country over investments in, and ownership of, each other's energy interests is an apt example. With petroleum growing increasingly scarce and costly and still being so vital, these flare-ups are not surprising. But we cannot afford to let them obscure the need to work together in response to a common problem, the reliance on OPEC oil. In the summer of 1981 a U.N. sponsored conference on new and renewable energy was held in Nairobi, Kenya. I would hope that session and ones like this conference will be just the beginning of a counter balance to OPEC.

Each country bases its decisions, foreign and domestic, on what it thinks is best. OPEC member Saudi Arabia, for example, is no exception. Sheik Yamani, Saudi Oil Minister, made the following comments regarding factors which determine oil production:

> "Whenever oil prices increase, large amounts of capital are invested in search of alternative sources of energy and in a search for oil in different areas. If we force Western countries to invest heavily in finding alternative sources of energy, they will. This would take no more than 7 to 10 years and would result in reducing dependence on oil as a source of energy to a point which will jeopardize Saudi Arabia's interests."

From time to time it has been proposed that the oil-consuming nations should form a petroleum purchasing cartel. Those ideas have never gained momentum because of the many practical problems rooted in self-interest.

Petroleum might be the world's dominant fuel right now but the potential for alternatives -- methane, alcohol, wind and solar, to name a few -- is immense in the future. We can speed up the process by making sure that we don't each reinvent the wheel but that we share energy technologies. Here those of us who are new energy consumers can work together. We can work together to even the world energy balance with our International Consortium for Alternative Fuels (ICAF).

This ICAF would not be faced with in-fighting among participants for an edge in terms of getting the lowest possible price. Instead, there would be an incentive to cooperate, to share technological advances in the use of alternative energy sources, and to adapt technologies to meet participating nations; special needs and circumstances.

Effective cooperation will show OPEC that energy stability lies in diversity and adaptation, not in reliance on a one-fuel policy. Methane clearly has a role in that diversity and this conference shows that we methane advocats recognize the advantage to be gained from working together.

Churchill once said, "It is not enough that we do our best, sometimes we have to do what is required." And I think that in the area of energy self-sufficiency we and the entire world, but particularly we in North America, can do what is required.

THE CANADIAN PROGRAM

Ray Perrault

Leader of the Government
Canadian Senate,
Ottawa, Ontario

Canadians are in a position to perceive the energy future of the world from a position of strength and not from weakness, and we are grateful for this. We produce more energy than we consume in this country. Compared to most other countries of the world, we are less vulnerable to the caprice of an international oil cartel and we are better able than most to break that dependency: our energy options are wide enough to preclude any sudden rush to energy choices that may compromise our social and environmental goals.

Our first goal is to wipe out net oil imports, within a decade, in this country. In moving towards this energy management objective, it will be essential to consider all realistic near-term options for the production, use and distribution of transportation fuels. Liquid transportation fuels comprise the one area where we believe opportunities for getting off oil are limited in the near-term. To lessen the transport sector's dependence on imported oil, the Canadian government has embarked upon a diversified approach, comprising four main elements, to the problem. Firstly, there is the stimulation of conservation through, for example, the recently promulgated Canadian fuel economy standards which provide vehicle manufacturers with fuel economy guidelines appropriate to the conditions which exist in this country. Secondly, the off oil program is aimed at reducing the proportion of oil used in each of the residential, commercial and industrial sectors in every province to more than 10% of the total energy, thus freeing oil for transportation use. Thirdly, an attempt to increase domestic production of oil from a variety of domestic sources including tar sands, heavy oil and improved recovery techniques. As well, this approach includes changing the way we refine and use fuels from current sources. For example, plans to upgrade 155,000 barrels a day of heavy oil to lighter products is one of the most efficient ways for this nation to obtain its oil product requirements. Fourthly, the development of new sources of fuel - for example, alcohol fuels, liquids from coal, fuel from synthesis gas, hydrogen and natural gas in its varied forms (compressed natural gas, liquid natural gas), methanol or gasoline produced from methanol.

The use of natural gas as a transportation fuel is of great interest in this context and British Columbia is to be commended for taking the initiative in organizing this conference. It should contribute to the formulation of policies which will enable methane to find an appropriate place in the Canadian transportation fuels supply equation.

What about natural gas as a vehicle fuel? A quick look at natural gas shows numerous benefits to its use in vehicles. The position of our national government on this subject is that firstly, there is an adequate resource base. Gas is plentiful in Canada, relative to oil. Over the past five years additions to gas reserves in conventional areas alone exceeded cumulative production by some 7.5 trillion cubic feet. So there is a broad concensus that this trend can be sustained for some time.

Secondly, we believe that methane is an excellent internal combustion engine fuel. It has an octane number of 130. It is clean burning so a longer engine life can be anticipated. Being a gas at all operating temperature ranges likely to be encounterd in Canada, it should give much easier cold-weather starting and better fuel economy while warming up. The low temperatures in Canada are not an insignificant consideration.

We have much to look forward to with the advancement of technology. Part of the challenge is learning how best to utilize that technology when it is viable.

Thirdly, methane as a transport fuel is environmentally attractive. Its combustion in engines causes considerably less hydrocarbon pollution than gasoline or diesel fuel so that methane fuel use would improve air quality relative to conventional liquid fuel.

And fourthly, it is our view as a national government that methane also appears, at first glance, to be an economically attractive fuel since the raw material cost in Canada and, indeed, worldwide tends to be lower than oil on an equivalent energy basis. Given such benefits, it may appear surprising that methane has not already played a major role as a transport fuel in Canada. Some of the difficulties associated with this fuel option are revealed in other chapters in this volume.

Firstly, current technology may appear somewhat unattractive for certain motor fuel applications when compared to conventional alternatives. Because of methane's low energy density, a vehicle powered by compressed natural gas loses storage space to the bulky tanks and fittings. The vehicle range tends to be low. The tank weight restricts the payload. Power loss occurs, especially when operating in the dual fuel mode. However, we are all aware that considerable ingenuity is being applied to overcome some of these technical difficulties.

Secondly, while much of settled Canada is now served by a natural gas distribution system, this gas must be compressed for use by vehicles. There are some 22,000 service stations in Canada serving Canada's population. Some ten percent of these offer diesel fuel, which is only now beginning to penetrate the vehicle market beyond the truck fleets. A very few of them now offer propane, another new vehicle fuel still in its infancy as far as this nation is concerned. None of them offer compressed natural gas. The task of creating a compressed natural gas refueling infrastructure is a large one even if already existing retail sites can be used. This will develop only slowly as equipment becomes available and as increasing demand for the fuel justifies the considerable risk of the capital investment required to dispense it.

Thirdly, safety codes for the utilization of compressed natural gas are only now being developed in Canada. It is encouraging that a draft preliminary standard for CNG components, as well as a draft supplement for the installation of CNG equipment in vehicles and the requirements for CNG refueling stations, has been produced by the Canadian Gas Association committee charged with this important task. Now these standards must be accepted by the respective provincial authorities so that they can be adopted. This can take a number of months to happen but it is coming. Beyond that is the question of the receptivity of municipal or local government authorities to the introduction of a new and unknown fuel. Local zoning regulations are likely to be impediments to the fuel distribution system expansion and this fact must be recognized.

A final difficulty, in our view, is the lack of knowledge and experience in Canada relating to natural gas as a vehicle fuel. This affects vehicle owners as potential users, the industry that must supply fuel, vehicles and equipment, and governments that must play a major role in any departure from current practice such as would be represented by widespread compressed natural gas use.

What is the approach of the national government to all of the problems and the challenges which I have outlined? Given the array of problems confronting the introduction of compressed natural gas in the face of some evident benefits, the Government of Canada is taking a measured approach to CNG. Our most important concern at this time is to seek potential solutions to the technical problems associated with compressed natural gas.

As part of the energy research and development funds announced in its National Energy Program, the government has solicited proposals for work on compressed natural gas. We are very pleased that the B.C. Research Council submitted a comprehensive plan to undertake and manage the research needed to support the development of better compressed natural gas technology in the eight areas described below.

Firstly, to evaluate, assess, modify and develop conversion hardware for a range of North American automotive power plants. Secondly, to evaluate and, if appropriate, develop light-weight CNG storage tanks sized to achieve 150 to 200 miles vehicle range and meet evolving safety standards.

Thirdly, to undertake the technical and commercial evaluation and, if appropriate, the development of compressors suitable for delivering natural gas to motor vehicle tank systems, again within the constraints of evolving safety standards. Fourthly, to undertake the theoretical and practical evaluation and development of materials and techniques which would enhance the high-density storage of natural gas without resorting to excessive pressures or cryogenic temperatures.

Fifthly, to establish a technical consultative service for safety regulators such that all the relevant safety aspects of evaluation and technical activities are made available to the regulators in a timely fashion.

Sixthly, to look at marketing. In that field B.C. Research will report on the current situation within developing and potential CNG markets and will suggest means of enhancing the market penetration of CNG both regionally and nationally.

Seventhly, to identify and evaluate systems for the retail marketing of CNG in accordance with current trade rquirements.

Lastly, to consider engine control system strategies: to evaluate, develop and provide information relative to equipment and systems for optimum control of sole or dual gasoline and CNG fuel automotive engines typical of those which could be retrofitted within the next ten years.

As you can see, this research program should provide answers to a number of the current technical questions. We expect the work will take five years to complete. The national government will direct approximately $6 million to this research alone.

Concurrent with the technical work, the Government of Canada is soliciting participants for large-scale demonstration programs in selected fleets and geographical areas. The objectives of these demonstration programs are to provide economic information on CNG as a motor fuel in commercial use leading to a better assessment of where are the best target markets for application; to provide a baseline of operation experience for tracking known and potential problems which need to be resolved to permit the rapid commercialization of the fuel; and to improve the opportunity for Canadian industry to develop expertise in the fields of conversion technology and equipment design. Ultimately it is hoped this will lead to the establishment of manufacturing facilities in Canada of CNG carburetion equipment. Finally, it is anticipated that demonstration programs will raise the visibility of this new fuel option among vehicle operators and manufacturers in Canada.

To achieve this objective, the Government of Canada is offering a cash grant of $600 per vehicle to individuals or fleet operators who convert their cars to operate on CNG and agree to provide data on their experience after a year of service.

The experience of other countries with CNG carburetion will provide some useful guidelines for Canadian policy formulation. It must be recognized, however, that a number of nations (for example, Italy, New Zealand and the United States) all have different political, economic and resource bases than Canada. What is effective in one country may not necessarily be transferable to Canada for this reason. We in the Federal Government will be paying attention to this experience for the guidance it might offer on the need for, and effect of, national policy instruments in promoting CNG.

In conclusion, the key concern of federal transport fuels policy is to ensure supplies in the future. CNG may be one answer but no single alternative fuel now in view can be expected to move in rapidly to fill the gap between our demand and the future supply of conventional products, including gasoline and diesel.

ENERGY POLICIES IN BRITISH COLUMBIA

Robert McClelland

Minister of Energy, Mines and Petroleum Resources

Government of British Columbia

Rather than discuss the scientific intricacies of methane as covered in other chapters, I would like to describe the energy policies being carried out by the Government of British Columbia. I do not believe that our society has fully arrived at a state of energy consciousness, even though the words themselves have become part of almost everyone's vocabulary. Even now, eight years after OPEC first set off the alarm clock, North America has not quite woken up to the realities and the opportunities of the energy challenge. But we are waking up fast -- and at least we are no longer wrapped in dreams of abundant cheap energy resources gained without effort or sacrifice.

Since the 1973 OPEC crisis, we in North America have developed what we might call "Stage One" energy consciousness. We have sat up and rubbed the sleep out of our eyes, and realized that we are into some kind of a new day. So, we have stopped buying gas guzzling V8's. We have started turning down the thermostat, or shutting off the lights when we leave a room. We are converting from oil to less costly and more available fuels like methane or wood waste. And, for most people, that constitutes energy consciousness.

But, as the shockwaves from OPEC continue to ripple through our society, we are finally getting shaken out of bed for good. We are entering "Stage Two" energy consciousness when we realize that there is a lot more to living in this oil-short world than saving a few gallons of gas. North American society will have become truly energy conscious when it becomes a matter of second nature to recognize the energy component built into every product, every action and every decision we make.

In Canada in recent years, energy issues have invaded the sphere of politics to such an extent that those matters not wholly displaced by energy problems have been irreversibly coloured by them. All our economic development programs, all our social programs, and certainly the area of intergovernmental relations have been altered -- sometimes to a drastic degree -- by the realities of life in the post-OPEC world.

It is therefore understandable if Canadians -- who see their elected representatives so immersed in energy matters -- might gain the impression that energy is essentially a problem of governments. But it is not. I can think of no other problem that so directly and so all-pervasively affects each of us in every aspect of our lives. And if the challenge is so universal, then society must respond with no less than a marshalling of each individual's efforts in support of the common good.

That universal response will be possible only when true energy consciousness becomes the social norm. We will have made our adjustment to OPEC when North Americans are as comfortably familiar with the balance between energy supply and demand as they are with the offensive and defensive capabilities of their favourite football team. Already, ideas about conservation and energy substitution that ten years ago seemed novel if not outlandish, are today's simple commonsense. Already, we take it for granted that energy considerations are paramount in planning for the future.

We are, all of us, becoming accustomed to thinking in new ways. Energy consciousness is becoming the background against which we view the world. It has become automatic, for every government or private company, to weigh the energy implications of any proposed course of action. As this new way of thinking continues to percolate through society, it will become automatic for everyone. By the end of this century, energy perspectives that have now flowed from the academic world into boardrooms and cabinet chambers will have seeped down to the primary school-yard. Children will find the facts of energy as obvious as the rules of baseball.

That vision of the future may seem a little too rosy for some, but consider how many incredible things we've come to take for granted in the past eight years. We've all had to become somewhat like the Queen of Hearts in _Alice in Wonderland_, who trained herself to believe six impossible things before breakfast. Energy consciousness -- real energy consciousness -- _is_ developing among North Americans, and it _will_ become second nature to us in time. We will become fully awake to the challenge of living without cheap oil, no matter how rude the awakening may be.

Governments, however, cannot fulfill their responsibilities

merely by trusting in the future. To paraphrase an old military slogan: we handle the immediate difficulty; the eventual takes a little longer. The role of government in helping a society to adjust to fundamental change is limited; as it should be. Governments do best when they don't try to do too much.

Legislatures may legislate, regulators may regulate, but no government can or should assume the citizen's responsibility to participate. Government must guide the citizen along the road towards energy security. In some places, government must smooth the way and level the worst obstacles. For those few citizens who simply cannot make their own way ahead, government must see that they are not left to fall by the wayside. But government cannot absolve the capable citizen of his duty to haul his own freight. We cannot lead if we are simultaneously trying to give a piggy-back ride to the able-bodied.

In this province, the government has established a set of energy policies and initiatives designed to lead to an energy-secure British Columbia -- and, by extension, an energy-secure Canada. We believe that our energy policies reflect both stage-one and stage-two energy consciousness; they respond both to our immediate difficulties and to the eventual promise held out to British Columbians by the substantial energy resources available to us.

We in British Columbia found ourselves in a curious energy situation after OPEC. We were, like most other developed economies in the western world, heavily dependent on oil. And only one quarter of the oil we burned was produced within our provincial borders. At the same time, however, we were and are in a region rich in almost every other form of energy resource, from the mundane to the futuristic.

We have large reserves of natural gas, large enough that the government has had to undertake a program to select from among the many proposals for gas development the ones which will most benefit British Columbia.

We have vast forests to produce biomass, and already our forest industry is able to displace almost 50 per cent of its bunker C oil requirements by burning hog fuel for heating and power generation.

We have over a billion tons of recoverable coal reserves, much of it thermal coal, in a decade when the world's demand for coal imports will more than double.

We have a province-wide hydroelectric system in place and expanding, and the potential for low-head hydro generation throughout

British Columbia.

We have natural hot-springs that may offer us geothermal power; and if tidal power generation ever becomes feasible, we have thousands of miles of fjord coastline.

We have coastal regions where constant winds may some day power electrical generators, and -- although the weather here in Vancouver sometimes seems to argue against it -- we are already beginning to tap the sun.

British Columbia even has mineable deposits of uranium ore, although nuclear development is not foreseen as one of our energy options and the government has placed a seven-year moratorium on uranium mining.

But, with all this energy at our command, we experienced the same shocks as most of the western world in the 1970's. We had relied too long on cheap and plentiful imported oil. And we had never efficiently organized all the various parts of our energy mosaic into one comprehensive design. Like most other western world governments, we have now rectified that oversight. Necessity is not only the mother of invention: she can also give birth to wise planning.

In February of last year, after a good deal of consultation and internal development, the provincial government brought forward a set of comprehensive energy policies for British Columbia. The aim is to make British Columbia energy secure. We want British Columbia to have sufficient energy to meet our needs: at all times, at fair prices, and with minimum vulnerability to forces outside our borders.

The government considered that our policies should be based on eight key elements. The first of these was to get our own act together. Various agencies and departments of the government had responsibilities that impinged to a greater or lesser degree on our energy situation. There was duplication and overlapping effort, and also areas where energy matters fell through the gaps between different arms of the administration. We unified responsibility for energy in British Columbia under one ministry, including the crown corporations that had charge of hydro and petroleum resources.

The second key element stems from simple commonsense. Since we are rich in all forms of energy other than oil, we should build on our strengths and cut our weaknesses. We are encouraging further development of our coal and natural gas industries. By 1984, the last major region of the province now dependent on oil for heating will be able to switch to natural gas, when the gas distribution system is extended to Vancouver Island. By 1995, this project will be cutting four million barrels of oil from our annual consumption total.

The third key element in our policies is conservation, particularly conservation of imported oil. Every gallon of gasoline that is not burned is another firm step towards energy security. And we practise what we preach -- we have already down-sized most of the government's auto fleet to more fuel-efficient vehicles. A federal-provincial study recently showed that compressed natural gas is the best alternative fuel available to us, and we have followed up that finding with a pilot program to use CNG in 50 of B.C. Hydro's transport fleet.

The program has now been extended and the Provincial Government is offering individuals $200 each for converting their vehicle to CNG, has removed the road tax on methane and has put in place a training program for mechanics at the Pacific Vocational Institute in Burnaby, B.C.

We are also practising conservation on a much broader scale. Officials of my ministry are just putting the finishing touches on a comprehensive internal conservation program for all government operations. We believe this system will be useful as a model for other public institutions and private enterprises. We also lend out our expertise in a more direct manner -- our conservation personnel travel the province helping industries, hospitals and schools to cut back on energy waste.

All of that fits into what I earlier described as "Stage One" energy consciousness. Here is an example of "Stage Two" thinking. It is now mandatory during the process of creating new legislation or regulations -- in any sector of the government -- to include an energy-impact analysis. Public servants are required to identify and justify the energy component of each government initiative before it is approved. Energy is never an afterthought -- it is now an integral part of our thinking.

Of course, we are doing more than just thinking. We have established an Energy Development Agency that launches or supports research into potential alternative energy sources, and into potential new uses for our traditional energy supplies. This conference itself is part of the government's resolve to examine all the options open to British Columbia.

Three very closely related elements of our energy policies are to streamline the process of reviewing and approving projects; to manage surplus energy reserves and exports; and to encourage energy development in the over-all interest of the province. As an example of how these elements tie together, I would like to outline briefly how we are setting about allocating surplus natural gas reserves.

By a conservative estimate, British Columbia has a gas surplus in the range of 50 to 75 billion cubic feet per year. We are now in the process of establishing criteria that will tell us more exactly what the available surplus is. In the past couple of years, a number of private companies and consortia have expressed serious interest in using that surplus gas in a variety of enterprises. Some want to establish liquid natural gas export plants. Others are interested in methanol or petrochemical production. Still others want to build fertilizer factories.

In June, 1981, I invited all interested parties to submit applications to the Provincial Government. To date, we have received a total of 15 proposals. They are now being analyzed by the government to determine which appear to offer the best net benefits to British Columbia. In the spring of 1982, those proposals which offer the most compelling benefits will be reviewed publicly by the Utilities Commission to detail the economic, social and environmental impacts of each.

When the review process has set the conditions for developing these energy projects, they will receive the appropriate certificates from the government. Without prejudging any of the applications, I can say that many of these natural gas proposals offer whole new energy industries for our province. And, in fact, some of the concepts being talked about at this conference may be present and in working order in British Columbia within the next few years.

The final element of our energy policies is the pricing of energy. The intricacies of establishing gas and oil prices in Canada would make an entire speech by themselves -- and a long one at that. Conference delegates who have come here from outside Canada may not be aware of the federal-provincial disagreements that have sometimes raged over questions of jurisdiction and taxation regimes. Canadian delegates, and particularly western Canadian delegates, are all too aware.

British Columbia and the Federal Government recently signed an energy agreement. Under this agreement the federal excise tax on our exported natural gas has been eliminated. The removal of the excise tax will take effect on October 1, 1981, and will restore over $500 million in revenues to the province over the life of the agreement.

The Federal Government has agreed to pay and administer all costs of the Petroleum Incentive Program (PIP). The federal assumption of all PIP costs will save the province about $600 million, the estimated amount the province would have had to pay in incentives had the terms of the Alberta/Ottawa agreement been followed.

Both governments recognize the importance of resolving the question of jurisdiction and control over our offshore resources and have undertaken to resolve these issues by early 1982.

The agreement also calls for the British Columbia government to pay to the federal government withheld taxes and interest collected by B.C. Hydro and the B.C. Petroleum Corporation.

I was most pleased that we could come to an agreement so quickly. We have taken a major step forward in helping British Columbia and the rest of Canada reach the goal of energy self-sufficiency and security.

Considering the advantages this province enjoys in the ownership of a multitude of energy resources, there is no reason why we should not become energy secure. Some thoughtful planning, some attention to the possibilities offered by innovative approaches, combined with a determination to do the best we can for our province -- all these together will bring us to our goal.

In 1973, we found out the hard way that we had left our energy security to chance and to the kindness of strangers. Now we are leaving no potential unexamined. I have mentioned already the conversion to biomass by the forest industry, and the broadening of natural gas distribution and development. We are investigating coal liquefaction. On Vancouver Island, a wind-powered generator is drawing electricity from the Pacific winds. In the Kootenays, a pilot geothermal plant is pumping power out of the earth's inner heat.

And we are looking seriously at the potential of methane as an alternative fuel. Despite all the conversions and conservation, we still rely on imported oil for 50 per cent of our total energy consumption. Along our highways, trucks still haul most of the freight moved between our scattered cities and towns. And that makes us vulnerable. A flare-up in the Persian Gulf, a re-unification of OPEC, even an intergovernmental dispute within Canada could have devastating effects. There are towns in British Columbia that are dependent for their very existence on their asphalt lifeline. Let the trucks stop rolling for just a few days, and there would be shortages of the essentials of life. Let the lifeline be cut for a week or more, and we would have pockets of disaster all across the province: empty food stores, and hungry people.

Today with a world oil glut, disaffection among member nations of the OPEC cartel, and a movement towards agreement between Ottawa and the provinces, this scenario of disaster seems unlikely, perhaps even farfetched. But no less farfetched than an international energy blockade seemed ten years ago. We've learned our lesson, and we won't trust so readily in the future again. In the past decade, the future has too often turned out to be disappointing and downright uncomfortable.

We must find a way in the next few years of cutting our transportation sector's dependence on imported oil. It may well be that methane is the answer. If so, this conference will not be an academic exercise. It will be a very crucial step on the way to keeping our outlying towns, our industries and our people safe from the worst of the world's energy storms.

APPENDICES

CONVERSION FACTORS

Approximate Conversion Factors

FOR CRUDE OIL *

FROM \ INTO	Tonnes	Barrels	Kiloliters (cub. metres)	1000 Gallons (Imp)	1000 Gallons (US)
	MULTIPLY BY				
Tonnes †	1	7.33	1.6	0.256	0.308
Barrels	0.136	1	0.159	0.035	0.042
Kilolitres (cub. metre)	0.863	6.29	1	0.220	0.264
1000 Gallons (Imp)	3.91	28.6	4.55	1	1.201
1000 Gallons (US)	3.25	23.8	3.79	0.833	1

* Based on world average gravity (excluding Natural Gas Liquids)
† One metric tonne = 1.1 tons one kg = 2.2 lbs.

MEASUREMENTS

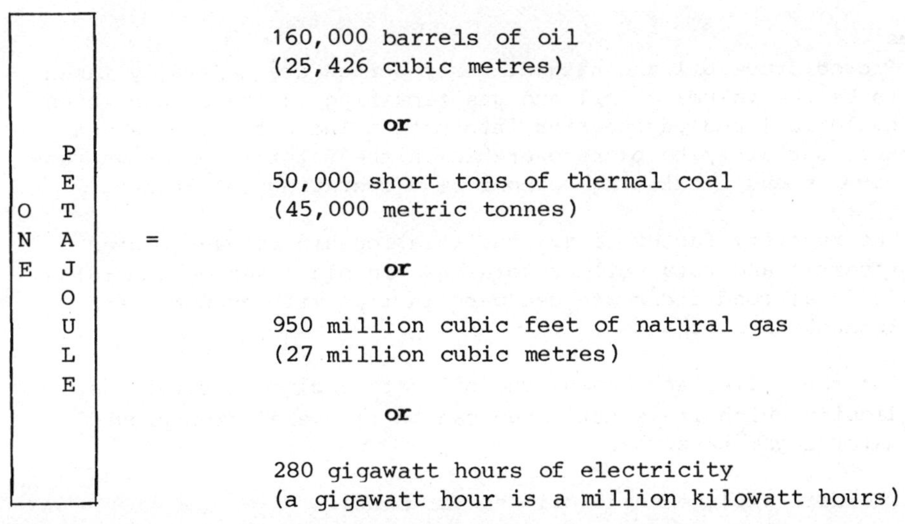

ONE PETAJOULE =

160,000 barrels of oil
(25,426 cubic metres)

or

50,000 short tons of thermal coal
(45,000 metric tonnes)

or

950 million cubic feet of natural gas
(27 million cubic metres)

or

280 gigawatt hours of electricity
(a gigawatt hour is a million kilowatt hours)

WORLD "PUBLISHED PROVED" RESERVES (AT END 1979)

Country/Area	OIL Thousand Million Tonnes	OIL Share of Total	OIL Thousand Million Barrels	NATURAL GAS Trillion Cubic Feet	NATURAL GAS Share of Total	NATURAL GAS Trillion Cubic Metres
U.S.A.	4.2	5.0%	32.7	194.9	7.6%	5.5
Canada	1.1	1.3%	8.1	88.1	3.4%	2.5
TOTAL NORTH AMERICA	5.3	6.3%	40.8	283.0	11.0%	8.0
Latin America	7.9	8.7%	56.5	144.5	5.6%	4.1
TOTAL WESTERN HEMISPHERE	13.2	15.0%	97.3	427.5	16.6%	12.1
Western Europe	3.2	3.6%	23.6	135.9	5.3%	3.8
Middle East	49.2	55.7%	361.8	739.8	28.7%	21.0
Africa	7.6	8.8%	57.1	210.4	8.1%	6.0
U.S.S.R.	9.1	10.3%	67.0	900.0	34.9%	25.5
Eastern Europe	0.4	0.5%	3.0	10.0	0.4%	0.3
China	2.7	3.1%	20.0	25.0	1.0%	0.7
Other Eastern Hemisphere	2.6	3.0%	19.4	128.2	5.0%	3.6
TOTAL EASTERN HEMISPHERE	74.8	85.0%	551.9	2 149.3	83.4%	60.9
World (excl U.S.S.R. E. Europe & China)	75.8	86.1%	559.2	1 641.8	63.7%	46.5
WORLD	**88.0**	**100.0%**	**649.2 ***	**2 576.8**	**100.0%**	**73.0†**

* Equivalent to 4.05 x 10^6 petajoules
† Equivalent to 2.70 x 10^6 petajoules

Source of data:

U.S.A. American Petroleum Institute and American Gas Association.

Canada. Canadian Petroleum Association

All other areas. Estimates published by the "Oil and Gas Journal" (Worldwide Oil issue - 31st December, 1979).

Notes:

1. Proved crude oil and natural gas reserves are generally taken to be the volume of oil and gas remaining in the ground which geological and engineering information indicate with reasonable certainty to be recoverable in the future from known reservoirs under existing economic and operating conditions.

2. The recovery factor, i.e. the relationship between proved reserves and total oil or total gas in place varies according to local conditions and can vary in time with economic and technological changes.

3. For the U.S.A. and Canada the oil data include natural gas liquids which it is estimated can be recovered from proved natural gas reserves.

4. The data exclude shale oil and tar sands.

5. Percentages are based on volume.

WORLD OIL PRODUCTION (1979)

Country/Area	Million Tonnes	Thousand Barrels Daily
U.S.A. Crude Oil	423.7	8 535
Natural Gas Liquids	59.4	1 675
Total	483.1	10 210
Canada	86.0	1 830
TOTAL NORTH AMERICA	**569.1**	**12 040**
LATIN AMERICA		
Argentina	24.5	470
Brazil	8.3	170
Columbia	7.5	150
Ecuador	10.5	215
Mexico	80.5	1 620
Trinidad	10.8	215
Venezuela	125.4	2 425
Other Latin America	12.5	260
TOTAL LATIN AMERICA	280.0	5 525
TOTAL WESTERN HEMISPHERE	**849.1**	**17 565**
WESTERN EUROPE		
Austria	1.7	35
France	1.2	25
Italy	1.8	30
Norway	18.8	385
Turkey	2.6	50
United Kingdom	77.9	1 600
West Germany	4.8	95
Yugoslavia	4.1	85
Other Western Europe	3.0	60
TOTAL WESTERN EUROPE	**115.9**	**2 365**
MIDDLE EAST		
Abu Dhabi	70.2	1 460
Dubai	17.6	355
Iran	155.6	3 125
Iraq	169.3	3 450
Kuwait	114.1	2 285
Neutral Zone	29.0	560
Oman	14.8	295
Qatar	24.6	505
Saudi Arabia	468.3	9 510
Sharjah	0.7	15
Other Middle East	12.2	235
TOTAL MIDDLE EAST	**1 076.4**	**21 795**

WORLD OIL PRODUCTION (1979)

Country/Area	Million Tonnes	Thousand Barrels Daily
AFRICA		
Algeria	56.4	1 215
Egypt	25.5	505
Libya	99.6	2 070
Other North Africa	5.4	115
Gabon	10.2	205
Nigeria	114.2	2 300
Other West Africa	13.1	260
TOTAL AFRICA	**324.4**	**6 670**
South Asia	14.9	300
SOUTH EAST ASIA		
Brunei	11.9	240
Indonesia	78.8	1 590
Other South East Asia	15.8	310
TOTAL SOUTH EAST ASIA	**106.5**	**2 140**
Japan	0.5	10
Australasia	21.9	455
U.S.S.R.	586.0	11 870
Eastern Europe	20.0	410
China	106.1	2 130
TOTAL EASTERN HEMISPHERE	**2 372.6**	**48 145**
WORLD (excl U.S.S.R. E. EUROPE & CHINA)	2 509.6	51 300
OF WHICH OPEC MEMBERS	1 544.5	31 285
WORLD	**3 221.7**	**65 710**

WORLD OIL CONSUMPTION (1979)

Country/Area	Million Tonnes	Percentage (Share of Total)	Thousand Barrels Daily
U.S.A.	862	27.7%	17,930
Canada	89.9	2.9%	1,895
Latin America	211.8	6.8%	4,395
TOTAL WESTERN HEMISPHERE	**1,164.6**	**37.4%**	**24,220**
Western Europe			
Austria	12.4	0.4%	250
Belgium & Luxembourg	29.4	0.9%	595
Denmark	16.1	0.5%	325
Finland	13.3	0.4%	265
France	118.1	3.8%	2,430
Greece	11.9	0.4%	240
Iceland	0.6		10
Republic of Ireland	6.2	0.2%	130
Italy	101.2	3.2%	2,045
Netherlands	38.5	1.2%	800
Norway	9.0	0.3%	185
Portugal	7.7	0.2%	160
Spain	47.3	1.5%	970
Sweden	28.4	0.9%	565
Switzerland	12.9	0.4%	270
Turkey	15.2	0.5%	310
United Kingdom	94.1	3.0%	1,935
West Germany	146.9	4.7%	3,045
Yugoslavia	15.8	0.5%	315
Cyprus/Gibralter Malta	1.5		25
TOTAL WESTERN EUROPE	**726.5**	**23.0%**	**14,870**
Middle East	74.8	2.4%	1,485
Africa	63.5	2.0%	1,310
South Asia	36.7	1.2%	755
South East Asia	116.9	3.7%	2,345
Japan	265.4	8.5%	5,495
Australasia	38.0	1.2%	800
U.S.S.R.	441.0	14.5%	8,925
Eastern Europe	101.1	3.2%	2,085
China	91.1	2.9%	1,835
TOTAL EASTERN HEMISPHERE	**1,955.0**	**62.6%**	**39,905**
WORLD (excl USSR E. EUROPE & CHINA)	2,486.4	79.4%	51,280
WORLD	**3,119.6**	**100%**	**64,125**

Differences between world production and consumption are accounted for by stock changes and oil "Destination not known".
* U.S. processing gains of 521,000 b/d in 1979 have been deducted from total domestic product demand.

CONTRIBUTORS

ANTAL, Michael

Professor, Mechanical and Aerospace Engineering
Princeton University

A PhD in applied mathematics from Harvard, Antal worked at the Los Alamos Scientific Laboratory as manager of the synthetic fuels from solid wastes project until 1975. At Princeton he worked on pyrolysis and in 1982 he joined the faculty of the University of Hawaii.

AXWORTHY, Robert

Vice-President, Marketing
Corken International Corporation
Oklahoma City, Oklahoma

Axworthy, who holds degrees in chemical engineering and pre-law from Washington State University, is a 17-year veteran of Corken, a leading manufacturer of pumps and compressors for the LPG industry. He is currently serving as chairman of the National LPG Association's Safety Committee.

BELLINI, Velio

Design Engineer
Nuovo Pignone
Florence, Italy

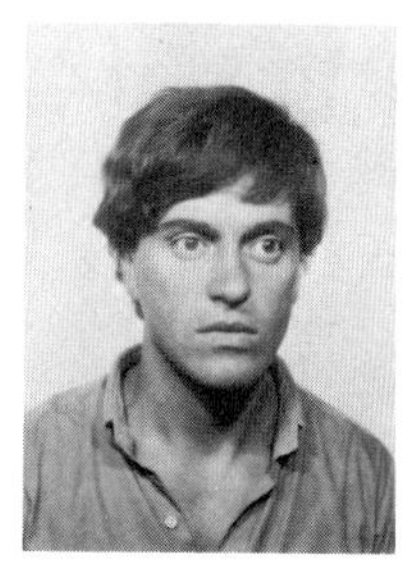

Bellini, 30, is involved with compressor design and engineering and has worked for Nuovo Pignone since 1970.

He received his degree in mechanical engineering from the Technical Institute of Florence.

BONVECCHIATO, Gustavo

Research and Development Branch
SNAM
Milan, Italy

Bonvecchiato, 52, graduated in mechanical engineering in Genoa. At SNAM he is involved in research and development of natural gas utilization and savings. His chapter was co-authored with Pietro Magistris, who is in charge of the CNG refuelling stations owned by SNAM.

BORN, Gerard

Alternate Fuels Laboratory
University of British Columbia
Vancouver, British Columbia

Born joined the UBC staff in 1980 but for 20 years prior to that he was at Princeton University in the Department of Mechanical and Aerospace Engineering.

His research has been connected with internal combustion, engine performance, engine parameter measurement and engine controls.

BOWEN, W.J.

Chairman, Chief Executive Officer
Transco Companies Incorporated
Houston, Texas

Starting with the Delhi Oil Co. in 1949, Bowen, 59, became CEO of Florida Gas Company, now Continental Resources Co., in 1960. He joined Transco in 1974.

Transco operates one of the nation's largest NG pipeline systems extending from Gulf Coast producing areas to East Coast states.

BROWN, Lloyd

Chief Executive Officer
Wellington Gas Company
New Zealand

Brown is currently President of the Gas Assoc. of New Zealand, a member of the NZ Energy Research and Development Committee and past chairman of the National Research Council. Brown has worked as an engineer in the U.S.A., the United Kingdom, Switzerland and Australia.

BUCHANAN, Hon. Judd

President
CNG Fuel Systems Limited
Calgary, Alberta, Canada

Member of Parliament from 1968 to 1980, Buchanan served as Minister of Indian and Northern Affairs, Minister of Public Works, and Minister of State for Science and Technology. As president of CNG Fuels since 1980, Buchanan is actively involved in the development of CNG as a vehicular fuel.

DEFFEYES, Kenneth

Professor of Geology
Princeton University

On faculty at Princeton since 1967 Deffeyes, 50, took his undergraduate degree in petroleum geology at the Colorado School of Mines. His research at Princeton has included plate tectonics, statistical estimation of oil and uranium resources and geochemistry. He has been a consultant to Exxon, Union Carbide and Ashland Oil.

DUNNING, W.C.

Managing Director
Caltex Oil Limited
Wellington, New Zealand

Born in Washington, DC, Dunning received a Masters in International Relations from George Washington University. He joined Caltex in 1957 and has held various marketing positions - including that of Chief Executive Officer with the Thailand and Phillipines companies.

DURBIN, Enoch J.

Professor of Mechanical and Aerospace Engineering
Princeton University

A graduate of Rennselaer Polytechnic Institute (PhD), Durbin has been on Princeton's faculty since 1965. In 1980/81 he headed the methane engine project at UBC's Alternate Fuels Laboratory. Durbin is a consultant to NATO and has been an advisor to many U.S. governmental agencies.

GLICKMAN, Congressman Dan

Congressman
United States House of
Representatives

Glickman, 37, has been the Democrat Congressman for the Wichita, Kansas area since 1976. He is chairman of the House Science and Technology Committee's subcommittee on Transportation, Aviation and Materials and sponsored the Methane Transportation Research, Development and Demonstration Act (1980).

GOLD, Thomas

Professor and Director
Center for Radio Physics
and Space Research
Cornell University

A Fellow of the Royal Society and a member of the U.S.National Academy of Sciences, Gold, 61, has been at Cornell since 1961. His work on earth outgassing of methane is well-known and he is the author of the explanation that pulsars are rotating neutron stars.

GRAHAM, Peter J.

Assistant Secretary (Oil & Gas)
Ministry of Energy
Government of New Zealand

With a background in industrial and economic development, Graham joined the Ministry of Energy in 1978. He is now charged with special responsibility for commercial development of alternative fuels.

KARIM, Ghazi A.

Professor
Department of Mech. Engineering
University of Calgary, Alberta

A graduate of the University of London, England, (PhD) in 1960, Karim has taught at London, Cambridge and Calgary. His current research projects include the utilization of hydrogen as a fuel, low temperature effects on combustion devices, air pollution and combustion of oil sands.

McCLELLAND, Honourable Robert

Minister of Energy, Mines and Petroleum Resources
Government of British Columbia

First elected as a Social Credit Member of Legislature in 1972, McClelland, 47, has held the portfolios of both Health and his current one. He is a member of the Government's Energy Development Agency, Chairman of the Cabinet Committee on Economic Development and is on the B.C. Hydro Board.

McELROY, R.O.

Manager, Fuels Program
B.C. Research
Vancouver, B.C., Canada

A PhD in hydrometallurgy from the University of B.C., McElroy has worked at B.C. Research since 1970. He is now responsible for management and planning of energy conversion projects including studies on coal liquefaction, fluidized bed combustion and methane as a transport fuel.

McGEER, Honourable Patrick L.

Minister of Universities, Science and Communications
Government of British Columbia

A PhD in chemistry from Princeton and an MD from the University of B.C., McGeer, 54, has been a Member of the Legislature since 1962. He has also held the portfolios of Education, Science and Technology. McGeer is on leave of absence from UBC where he heads the Neurological Science Division.

PERRAULT, Senator Ray

Leader of the Government
Canadian Senate
Ottawa, Canada

A Senator since 1973, Perrault is a member of the federal cabinet. He served as Liberal Member of the B.C. Legislature from 1960 to 1968 and as Liberal Member of Parliament for Burnaby-Seymour from 1968 to 1972.

PISCHINGER, Georg H.

Manager, Research Division
Volkswagen do Brasil
Sao Bernardo, Brazil

Receiving a PhD in engineering from the Technical University Graz, Austria, in 1973 (thesis: Friction losses in Diesel Engines), Pischinger worked until 1978 with Audi in Germany. At Volkswagen his applied research division dealt mainly with ethanol engines.

SIEKMANN, Rolf W.

Manager, Research Department
Volkswagen do Brasil
Sao Bernardo, Brazil

A chemistry graduate from the Technical University of Hanover, Germany, (1975), Siekmann worked as an assistant professor at the Technical University, Clausthal-Zellerfeld, Germany, and received his PhD from there in 1979 (thesis: Coal Liquefaction). He has been with Volkswagen since 1980.

TANAKA, Munenobu

Professor, Mechanical Engineering
Tokyo University of Agriculture and Technology

Tanaka was born in Kobe, Japan, in 1922. As Chairman of the Mechanical Engineering Department at his university, Tanaka's major area of research is internal combustion engines, especially their dynamics and inlet systems.

VAN DER WEIDE, Jouke

Research Institute for Road Vehicles (TNO)
Delft, The Netherlands

Now the head of the department of internal combustion engines at the Research Institute, van der Weide studied at the Technical University, Delft. He worked on LPG, methanol and natural gas, especially developing hardware for the application of alternative fuels to diesel engines.

WRIGHT, John E.

General Manager
Gas Service Energy Corporation
Kansas City, Missouri

Wright graduated from the University of Missouri in electrical engineering. Since 1969 he has been involved in the design and operation of natural gas vehicle systems. The Gas Service Energy Corp., manufactures and distributes, internationally, a complete line of NG vehicle conversion and refueling equipment.

CONFERENCE EXHIBITORS

ECKLUND, Eugene

Alternative Transportation Fuels Utilization
Department of Energy
United States Government

Ecklund, 61, has spent nearly 40 years in the electronics and aerospace industry. He has been with the U.S. Federal Government since 1974. Ecklund graduated from the Polytechnic Institute of New York.

LANGDON, Ray

Consultant
Energy Development Agency
Government of British Columbia

Consultant to the B.C. Government's methane project, Langdon, 59, is responsible for promotion of natural gas as a motor vehicle fuel. He has had 30 years experience with British American Oil and Gulf Oil Marketing in Western Canada

NAVARRE, Donald

Senior Vice-President, Marketing
Washington Natural Gas Company
Seattle, Washington

Past Chairman of the American Gas Association's marketing section and Chairman of the marketing services section of the Pacific Coast Gas Association, Navarre, 53, is a graduate of the University of Toledo's business school.

STEWART, Robert

Deputy Minister
Ministry of Universities, Science and Communications
Government of British Columbia

Stewart, 58, received his PhD in physics from Cambridge (1952). He has taught at several universities and was director-general of Canada's Institute of Ocean Sciences from 1970 to 1979. He is a member of the Royal Society and is on several international committees on climate research.

INDEX